T0142822

Springer Atmospheric Sciences

More information about this series at http://www.springer.com/series/10176

Boris M. Smirnov

Global Atmospheric Phenomena Involving Water

Water Circulation, Atmospheric Electricity, and the Greenhouse Effect

 Springer

Boris M. Smirnov
Joint Institute for High Temperatures
Russian Academy of Sciences
Moscow, Russia

ISSN 2194-5217 ISSN 2194-5225 (electronic)
Springer Atmospheric Sciences
ISBN 978-3-030-58041-4 ISBN 978-3-030-58039-1 (eBook)
https://doi.org/10.1007/978-3-030-58039-1

This Springer imprint is published by the registered company Springer Nature Switzerland AG
The registered company address is: Gewerbestrasse 11, 6330 Cham, Switzerland

Preface

The analysis of elementary processes involving atmospheric water is combined in this book with the contemporary understanding of global atmospheric problems which include water circulation through the atmosphere, atmospheric electricity, the greenhouse effect and other global atmospheric problems. Using simple models, as the model of standard atmosphere, allows one to understand deeper these problems and obtain the estimations for various parameters of global atmospheric processes. The basic attention we pay to condensed water that is concentrated in clouds which consists mostly of water microdroplets. In addition, accompanied atmospheric processes are analyzed. As a result, each atmospheric phenomenon is a chain of certain processes.

As one of these phenomena, water circulation consists in the evaporation of water from the Earth's surface, capture of water molecules by air vortices in the course of convection, and partial condensation of water at altitudes of several kilometers with formation of clouds, consisting of water microdroplets. Returning water to the Earth's surface proceeds both in the form of water molecules which partake in convection motion of air or by rain and water microdroplets.

Atmospheric electricity is the secondary phenomenon with respect to water circulation and the atmosphere is divided into two parts, where the first one contains cumulus clouds which include almost all condensed water in the atmosphere in the form of water microdroplets mostly. Water microdroplets are formed as a result of the mixing of wet air from near-surface atmospheric regions and cold layers at an altitude of a few kilometers. Simultaneously growing microdroplets are charged by attachment of molecular ions formed under the action of cosmic rays. These microdroplets become charged because of different mobilities of positive and negative molecular ions. Charged microdroplets fall down that leads to the separation of atmospheric charges and the creation of the atmospheric electric field. Because cumulus clouds consist of microdroplets with charges of the same sign, they have a high electric potential and may be a source of atmospheric breakdown in the form of lightning. So, charging of the Earth proceeds due to cumulus clouds, while its discharge results from the currents of molecular ions formed under the action of cosmic rays in regions of the clear-sky atmosphere.

The greenhouse effect results from atmospheric emission in the infrared spectrum range. Water is the main greenhouse component, such that approximately 50% of the atmospheric radiative flux toward the Earth is created by water molecules, and 30% of the radiative flux results from cloud emission. It is of importance that atmospheric emission to the Earth is separated toward the Earth and outside because of a large optical thickness of the atmosphere. In addition, emission of clouds and molecular components of the atmosphere are separated in regions of rare clouds which occupy the basic part of the atmosphere.

In this book, we consider the above global atmospheric problems involving water in detail.

Moscow, Russia   Boris M. Smirnov

Contents

Chapter 1
Introduction

The aim of this book is to study the global processes involving atmospheric water that is present in the atmosphere both in the form of free water molecules and as a condensed phase in the form of microdroplets and snow particles located mostly in cumulus clouds. The global atmospheric phenomena under consideration include water circulation in the atmosphere, atmospheric electricity, and greenhouse phenomenon. These phenomena are studied widely both by measurement of their parameters and on the basis of the theoretical description of these phenomena, as well as their aspects. The latter gives the physical picture of the phenomenon and allows one to describe its physical picture. Analyzing these phenomena in this book, we add to the macroscopic processes of each phenomenon some elementary processes involving water molecules and water microdroplets. Finally, this leads to a deeper description of some aspects of the phenomena under consideration.

In considering global phenomena, we are basing on atmosphere parameters averaged on both time and the globe. They are included in the model of standard atmosphere [1]. The analysis on the basis of this model gives a qualitative description of these phenomena. We start from the circulation of water in the atmosphere [2] according to which water evaporates from the Earth's surface in the form of water molecules and returns both in the form of molecules and water microdroplets or snow particles. But formation of water condensed phase, i.e. water microdroplets and snow particles, is possible only in a supersaturated water vapor because the thermodynamic equilibrium between free water molecules and water droplets is established fast in the atmosphere. Note that within the framework of the model of the standard atmosphere the air humidity is restricted at any altitude, i.e. conditions of condensation are not fulfilled. This means that water condensation in motionless air of the standard atmosphere is impossible. Hence, formation of the water condensed phase in the atmosphere results from the action of vertical winds which mix warm wet air from low atmosphere altitudes with cold air of high altitudes. From this, it follows that creation of a supersaturated water vapor in the atmosphere proceeds randomly

B. M. Smirnov, *Global Atmospheric Phenomena Involving Water*, Springer Atmospheric Sciences, https://doi.org/10.1007/978-3-030-58039-1_1

and the portion of the water condensed phase in the atmosphere is small compared with that of free water molecules. This acts on water fluxes toward the Earth.

In the course of the formation and growth of water droplets in the atmosphere, the equilibrium is supported between a water vapor consisting of free water molecules and condensed water, i.e. water microdroplets. At low droplet sizes, the dominant growth mechanism is the coalescence one. In this process, small microdroplets evaporate and large droplets grow in the competition of evaporation and attachment processes. As a result, the average size of the droplets increases. At large droplet sizes, the gravitational fall mechanism of droplet growth dominates. In this case, a large droplet overtakes a small one in the course of their falling, and they join at their contact. In clouds, these processes are combined with the charging of microdroplets as a result of ion attachment to them, and ions are formed in atmospheric ionization under the action of cosmic rays. Hence, the process of droplet growth occurs simultaneously with electrical processes in clouds consisting of water microdroplets. The growing microdroplets become charged, and this influences the growth process. This testifies the connection between processes of growth of water microdroplets in the atmosphere and atmospheric electricity.

From this it follows that atmospheric electricity is a secondary phenomenon with respect to water circulation in the atmosphere. This means that electric currents in the atmosphere result from atmospheric ions containing water molecules, as well as from charged microdroplets and microparticles consisting of water molecules. The key electric process in the atmosphere which characterizes the separation of the positive and negative charges in it is the falling down of negatively charged water microdroplets. This process leads to negative charging of the Earth, and its discharging proceeds through drift of positive and negative molecular ions in the clear sky under the action of the Earth's electric field.

Because the Earth is charged negatively, one can accept that falling water microdroplets are charged negatively, and their mass is enough high such that the gravitation force acted on the microdroplet (or its weight) exceeds the electric force acted on this droplet due to its charge. Note that for molecular ions, an inverse relation is fulfilled. In considering electric processes in the Earth's atmosphere, one can divide the atmosphere region into two parts. The first part of the atmosphere occupied a small sky area is covered with cumulus clouds, and processes of charging of the Earth's surface occur due to this atmosphere part. In the other atmospheric part, ions formed under the action of cosmic rays move in the electric field of the Earth, such that this part is responsible for atmosphere discharging. The sum of these processes is included in the so-called global electrical circuit [3, 4] which characterizes the Earth's atmosphere as an electric system.

The Earth as an electric system may be represented as a spherical capacitor with the Earth's surface as an electrode, and the other electrode relates to atmospheric layers at altitudes with a high conductivity. It is convenient to take the ionosphere as the upper atmospheric electrode, and the air conductivity at this electrode is high and exceeds that of the troposphere. The electric voltage between these electrodes is supported by formation of atmospheric ions under the action of cosmic rays, and a current due to these ions in the Earth's electric field leads to the Earth's discharging in regions of clear sky. The Earth is charged negatively, and its charge is formed by currents in regions of cumulus clouds as a result of gravitation falling of negatively charged microdroplets of water.

In the course of gravitation falling of charged water microdroplets which constitute cumulus clouds, they acquire a high electric potential and create an atmospheric electric field whose strength is lower two orders of magnitude compared to that caused by electric breakdown in dry air. Nevertheless, the electric voltage of clouds with respect to the Earth may exceed the average voltage between the Earth and the ionosphere in two-three orders of magnitude. Such an electric potential may cause a specific electric breakdown of the atmosphere in the form of lightning. Some aspects of physics of the lightning phenomenon are considered below.

In considering the greenhouse atmospheric phenomena, we continue the general line of this book which consists in applying the information about elementary processes with the participation of atoms, molecules, and particles with atmospheric phenomena. Then we consider the Earth's atmosphere as a weakly nonuniform flat gaseous layer and combine emission of various atmospheric components as it was made for outgoing radiation of the atmosphere [5–7] by using HITRAN bank data for radiative transitions of molecules [8, 9]. This approach joins with the model "line-by-line" [10] that implies the evaluation of radiative fluxes in the infrared frequency range at each frequency. This approach accounts for the optical interaction between various components that take into account the absorption of radiation created by a certain molecular component at a given frequency with other components. This fact is not taken into account in contemporary climatological codes [11, 12].

Basic greenhouse components of the Earth's atmosphere are H_2O and CO_2 molecules, as well as water microdroplets which constitute the clouds. In considering the atmospheric emission toward the Earth, we keep below the scheme [13] which along with the emission of molecular atmospheric components and a nonuniformity of the atmosphere takes into account emission and absorption by clouds. In addition, we attract the energetic balance of the Earth and its atmosphere that allows us to normalize the total radiative flux of the atmosphere in using the appropriate average altitude of cloud location. As a result, water molecules give the contribution of 51% to the total atmospheric radiative flux emission to the Earth. In addition, 29% of the total radiative flux is created by water microdroplets, 18% of the total radiative flux is due to CO_2 molecules on average [13], as well as approximately 2% of the total radiative flux relate to CH_4 and N_2O atmospheric molecules.

As is seen, water microdroplets of clouds are of importance for the emission of the atmosphere in the infrared spectrum range. But clouds have a non-regular structure and the water density in them may vary in wide limits. Therefore, it is impossible to use the average density of condensed water in the atmosphere which follows from the analysis of cumulus clouds in atmospheric electricity because the area of cumulus clouds over the Earth's surface is relatively small. The mass of atmospheric water in microdroplets which are responsible for cloud emission is one order of magnitude less than the water mass in the atmosphere.

Let us mark one more peculiarity of water cloud microdroplets which are radiators and absorbers simultaneously. The absorption coefficient of pure liquid water in the infrared spectrum range exceeds by approximately seven orders of magnitude that in the visible spectrum range. Correspondingly, the sky with microdroplets consisting of pure water may be transparent to the human eye, i.e. for visible radiation, and be an absorber for infrared radiation. This atmospheric property is known as the Twomey effect [14, 15].

The mean free path of photons for infrared spectrum range which corresponds to thermal emission of the Earth is $(3–30)\,\mu m$, and we evaluate the cross section for scattering of infrared radiation on liquid water microdroplets on the basis of the Mie theory [16]. In addition, if the total atmospheric water is transformed in the liquid state and is distributed uniformly over the globe, the thickness of the water layer is approximately 2.5 cm. From this, it follows that if the specific mass of condensed water in clouds exceeds 0.1% of the average water mass in the atmosphere, the optical thickness of a given atmospheric column with respect to infrared radiation exceeds one. Thus, clouds become opaque for infrared radiation if the mass of condensed water in a column of atmospheric air is less than the average one and that in cumulus clouds. Clouds with higher values of densities of condensed water absorb infrared radiation effectively. Note that such clouds become visible to the human eye only after the attachment of salt molecules or dust particles to microdroplets of clouds, if these admixtures are effective absorbers in the visible spectrum range.

Thus, in this book we consider three global atmosphere phenomena involving atmospheric water, namely, circulation of water through the atmosphere, atmospheric electricity, and the greenhouse phenomenon. The part of atmospheric electricity connected with the negative Earth's charging is the secondary phenomenon of the water circulation. Moreover, the Earth's charging proceeds due to atmospheric condensed water formed in the course of water circulation. This process occupies a small part of the area of the Earth's surface and influences weakly on other global phenomena. On the contrary, the greenhouse atmospheric phenomenon is separated from the above phenomena and does not depend on them. It proceeds over the all globe and uses a small part of condensed atmospheric water.

Note that in this consideration, we analyze global properties of the atmosphere, and hence yield numerical parameters have an estimated character. Therefore, our approach is qualitative and allows us to obtain the physical picture of atmospheric phenomena and the connection of parameters of these phenomena with those of elementary processes under consideration.

References

1. *U.S. Standard Atmosphere.* (Washington, U.S. Government Printing Office, 1976)
2. R. Braham, J. Meteorol. **9**, 227 (1952)
3. C.T.R. Wilson, J. Franklin Inst. **208**, 1 (1929)
4. E. Williams, Atmosp. Res. **91**, 140 (2009)

5. R.T. Pierrehumbert, *Principles of Planetary Climate* (Cambridge University Press, New York, 2010)
6. R.T. Pierrehumbert, Phys. Today **64**, 33 (2011)
7. W. Zhong, J.D. Haigh, Weather **68**, 100 (2013)
8. https://www.cfa.harvard.edu/
9. http://www.hitran.iao.ru/home
10. R.M. Goody, *Atmospheric Radiation: Theoretical Basis* (Oxford University Press, London, 1964)
11. https://www.ipcc.ch/report/ar4/wg1/climate-models-and-their-evaluation
12. Intergovernmental panel on climate change. Nature **501**, 297–298 (2013) http://www.ipcc.ch/pdf/assessment?report/ar5/wg1/WGIAR5-SPM-brochure-en.pdf
13. B.M. Smirnov, *Transport of Infrared Atmospheric Radiation.* (Fundamentals of the Greenhouse Phenomenon.) (de Gruyter, Berlin, 2020)
14. S. Twomey, Geofis. Pure Appl. **43**, 227 (1959)
15. S. Twomey, J. Atmos. Sci. **34**, 1149 (1977)
16. G. Mie, Annalen der Physik **330**, 377 (1908)

Chapter 2
Global Properties of the Earth's Atmosphere

Abstract Within the framework of the model of the standard atmosphere, general properties and processes are considered in atmospheric air involving water molecules and water microdroplets. The convective motion of air which includes air vortices and jets or wind, captured atmospheric molecules as well as small particles leads to separation of molecules from microparticles because of their higher inertia. The flux of atmospheric water to the Earth due to molecules is less than that due to water microdroplets. The thermal state of the Earth as a whole and the evolution of the global temperature are analyzed in this time and in the past. The problem of carbon dioxide in the atmosphere is represented. Energetics of the Earth and its atmosphere is analyzed.

2.1 Water in Atmosphere

2.1.1 Parameters of Standard Atmosphere

In considering the behavior of water molecules in atmospheric air, we are guided by physics of the atmosphere and its general properties which are represented in books [1–17]. In the description of various aspects of physics of atmospheric water, we will be basing on the general properties of atmospheric physics. One can take into account that approximately 75% of the total mass of atmospheric air and 99% of the total mass of atmospheric water are concentrated in the troposphere [18]. The depth of this atmospheric layer is ~ 10 km that is small compared to the Earth's radius 6370 km. Therefore, one can consider the troposphere as a flat air layer which parameters depend only on a distance h from the Earth's surface. This corresponds to the model of standard atmosphere [19] which is described by average parameters and will be used for the subsequent analysis.

Within the framework of the model of the standard atmosphere, the global temperature, i.e. the temperature near the Earth's surface, is equal to 288 K. Removal from the Earth's surface leads to a decrease in the air temperature because of air expansion. One can assume this expansion as the adiabatic one [20–23], i.e. an air volume which is displaced to another altitude, does not exchange by energy with the

© The Editor(s) (if applicable) and The Author(s), under exclusive license
to Springer Nature Switzerland AG 2020
B. M. Smirnov, *Global Atmospheric Phenomena Involving Water*,
Springer Atmospheric Sciences, https://doi.org/10.1007/978-3-030-58039-1_2

surrounding air. For pure air consisting of nitrogen and oxygen molecules, this leads
to the temperature gradient

$$\frac{dT}{dh} = -10\frac{K}{km},$$ (2.1)

in the troposphere. We represent below the atmospheric air to be consisting of identical molecules whose mass is 29 $a.u.m.$ In processes under consideration, air is a
buffer gas.

In reality, the temperature gradient is lower than that which follows from formula
(2.1). The reason is that a temperature decrease causes condensation of atmospheric
water. For the model of standard atmosphere [19] which approximates observed
parameters of a real atmosphere, the temperature gradient is equal to

$$\frac{dT}{dh} = -6.5\frac{K}{km}$$ (2.2)

A temperature decrease is finished at the tropopause, i.e. the boundary between the
troposphere and stratosphere that is located at altitudes between 11 and 20 km. At
larger altitudes the air temperature increases with an increasing altitude because of
the absorption of solar radiation by stratospheric ozone. We below restrict by the
analysis the troposphere where the temperature gradient is given by formula (2.2),
and the tropopause temperature is 217 K for the model of standard atmosphere .

In considering atmospheric air as a united system contained of gaseous molecules
with the mass of 29 $a.u.m.$, we approximate the altitude dependence on the number
density of atmospheric molecules as

$$N(h) = N_o \exp\left(-\frac{h}{\Lambda}\right),$$ (2.3)

where $N_o = 2.55 \times 10^{19}$ cm^{-3} is the number density of air molecules near the Earth's
surface, and the scale $\Lambda(0) = 8.4$ km, as it follows from the model of standard atmosphere [19] constructed on the basis of averaging of parameters of the real atmosphere.

We note that in reality, atmospheric parameters under consideration in some geographical points and at various times are different. Using averaged parameters of
the model of the standard atmosphere, we hence use the approximation that gives a
qualitative description of atmospheric processes.

2.1.2 Atmospheric Water Vapor

Water is the basic additional atmospheric component which partakes in the optical and
electrical processes of the atmosphere. Therefore, we first consider the parameters
of a water vapor which is located in the atmosphere and which consists mostly of

water molecules. An average amount of atmospheric water is 1.3×10^{19} g [24–27] that corresponds to the mean water density of $3\,g/m^3$ in the atmosphere. One can compare the water mass in the atmosphere with that of atmospheric air 5.1×10^{21} g which relates to nitrogen and oxygen molecules. This corresponds to the average concentration of water molecules in atmospheric air as approximately 0.4%, whereas near the Earth's surface the average concentration of water molecules in the air is equal to 1.7%.

The total rate of water evaporation from the Earth's surface is 4.8×10^{20} g/year [26, 28–31] or 1.5×10^{13} g/s $= 15 \times 10^6$ ton/s (only 1.0×10^{18} g/year is found in the form of snow), and the same rate relates to water returning to the Earth's surface. As it follows from this, the power of the evaporation process is equal to 2.4×10^{16} W (see Fig. 2.1) which is returned mostly as a result of water condensation in the atmosphere. In this manner, the above power is transferred from the Earth to the atmosphere.

Taking the ratio of the total amount of atmospheric water and the total rate of its evaporation from the Earth's surface, one can determine the average residence time of water molecules in the atmosphere as approximately 9 d [31]. Along with this, if we collect all atmospheric water, transform it into a liquid, and distribute uniformly all over the globe's surface, the height of this liquid layer would be 2.5 cm [32].

It should be noted that atmospheric water contains a small part of Earth's water whose mass is 1.4×10^{24} g. If Earth's water would be distributed over the Earth's surface uniformly, the layer thickness would be 2.7 km. From this, it follows that most part of this water is located in the underground. Note that 96% of Earth's water is salty. In addition, open water located on the Earth's surface is a source of atmospheric water and is found in equilibrium with it.

An equilibrium between atmospheric water and that located at the Earth's surface results from evaporation and precipitation processes. But the average specific rate of water evaporation from oceans is higher than that from the land because of higher air moisture above oceans. In order to compensate this difference, there is a flux from land to oceans, and the water balance at the Earth's surface is given in Fig. 2.1. As it follows from Fig. 2.1, mean water fluxes from the Earth's surface to the atmosphere and vice versa do not coincide for the land and oceans, such that the flux of water

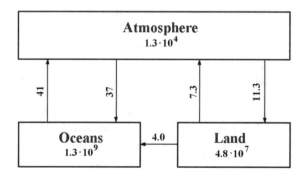

Fig. 2.1 Rates of exchange by water between land, oceans, and atmosphere which are expressed in 10^{18} g/year and are given near arrows. The water amount in each object is expressed in 10^{15} g (billion ton) and is indicated inside a corresponding rectangle [26, 28–31]

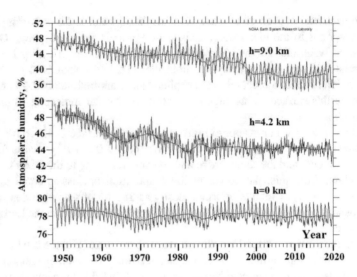

Fig. 2.2 Evolution of the global moisture of atmospheric air at indicated altitudes [33]

evaporation from the land is less than the precipitation water flux. Rivers which transfer water from the land to oceans compensate for the difference of these fluxes and provide the water balance.

One can characterize the concentration of atmospheric water molecules in air by its moisture or humidity η [33] which is introduced as

$$\eta = \frac{N(H_2O)}{N_{sat}(T)},\qquad(2.4)$$

where $N_{sat}(T)$ is the number density of water molecules at the saturated vapor pressure $p_{sat}(T)$ at a given temperature, i.e. the number density of water molecules at a given temperature if this temperature becomes the dew point [34]. At this temperature, the partial pressure of a water vapor is equal to the saturated vapor pressure at this temperature. Figure 2.2 contains evolution of the average moisture in time. Note that altitudes considered in Fig. 2.2 relate to the air pressures of 1, 0.6 and 0.3 atm, and this allows one to compare the altitude dependence for molecules of atmospheric air and water molecules. In addition, according to Fig. 2.2, large fluctuations accompany the average number density of water molecules.

The altitude dependence for the number density of water molecules, as well as this dependence for the air moisture, allows one to find parameters for the altitude dependence of the number density of water molecules. In particular, from Fig. 2.2 it follows that the concentration of water molecules in atmospheric air is 1.7, 0.34 and 0.045% above the Earth's surface at altitudes of 4.2 and 9 km correspondingly according to averaging of data in 1950–1980, and these values according to averaging for 1980–2010 are correspondingly 1.7, 0.33, and 0.039%. It is convenient to approximate the number density N_w of water molecules as an altitude function in a simple form

$$N_w(h) = N_o \exp(-h/\lambda), \tag{2.5}$$

where h is the altitude. Then from 1950 to 1970 data for the moisture (Fig. 2.2 and [36–38]), one can find the following parameters of formula (2.5) as $N_o = 4.2 \times 10^{17}\,\mathrm{cm}^{-3}$, $\lambda = (2.0 \pm 0.1)\,\mathrm{km}$, and for 2000–2020 we have $N_o = 4.3 \times 10^{17}\,\mathrm{cm}^{-3}$, $\lambda = (1.9 \pm 0.1)\,\mathrm{km}$. As is seen, these values coincide within the limits of their accuracy, and we take as the parameters of formula (2.5) the following values:

$$N_o = 4.3 \times 10^{17}\,\mathrm{cm}^{-3}, \quad \lambda = 2.0\,\mathrm{km} \tag{2.6}$$

One can reveal some contradiction for parameters of formula (2.6). Indeed, N_o is the average number density of water molecules at the Earth's surface, and if we take this value according to formula (2.6) and the global temperature $T = 288\,\mathrm{K}$, one can obtain the water moisture at the Earth's surface as $\eta = 100\%$. According to data of Fig. 2.2, the average air moisture near the Earth's surface is approximately 80%. A difference of these values obtained on the basis of indicated operations may be connected with the accuracy of these operations.

From this, one can find the total amount of atmospheric water, and the mass of atmospheric water is equal to

$$M_w = N_o \lambda S m_w, \tag{2.7}$$

where $S = 5.1 \times 10^{18}\,\mathrm{cm}^2$ is the area of the Earth's surface, and $m_w = 3.0 \times 10^{-23}\,\mathrm{g}$ is the mass of an individual water molecule. From this formula, it follows that $M_w = 1.3 \times 10^{19}\,\mathrm{g}$ that coincides with the above value. Thus, with an accuracy of a few percent, the total amount of atmospheric water is identical, being obtained in different ways. Thus, the average number density of water molecules near the Earth's surface is $N(H_2O) = 4.3 \times 10^{17}\,\mathrm{cm}^{-3}$, the number density per unit vertical column is equal to $n(H_2O) = 1.4 \times 10^{23}\,\mathrm{cm}^{-2}$, and its variation from the end of nineteenth century is $\Delta n(H_2O) = 5.7 \times 10^{21}\,\mathrm{cm}^{-2}$, if we take in accordance with [39] that an increase of the amount of atmospheric water vapor is 4%. As is seen, variations of an amount of atmospheric water exceed several times that of carbon dioxide.

One can introduce also the concentration of water molecules in air as

$$c(h) = \frac{N_w(h)}{N(h)} = c(0) \exp\left(-\frac{h}{\lambda_*}\right), \quad \frac{1}{\lambda_*} = \frac{1}{\lambda} - \frac{1}{\Lambda} \tag{2.8}$$

Since $\Lambda = 8.4\,\mathrm{km}$, $\lambda = 2.0\,\mathrm{km}$, one can obtain $\lambda_* = 2.6\,\mathrm{km}$. In addition, $c(0) = 1.7\%$ within the framework of the model of standard atmosphere.

The altitude distribution for the number density of atmospheric water molecules may be determined on the basis of balloon measurements [40, 41]. In this case, a balloon rises up to the stratosphere and then goes down. In the course of this, the number density of water molecules is measured in atmospheric air on the basis of

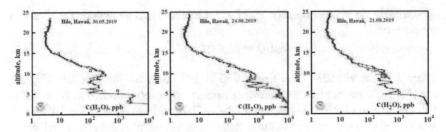

Fig. 2.3 Concentration of water molecules in the atmosphere as a result of balloon measurements at Hillo (Hawaii) [35] both during ascent and descent. Open squares correspond to formula (2.5)

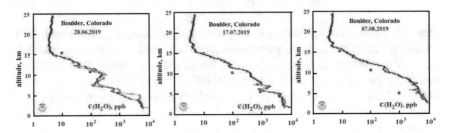

Fig. 2.4 Concentration of water molecules in the atmosphere as a result of balloon measurements at Boulder (Colorado, USA) [35] both during ascent and descent. Open squares correspond to formula (2.5)

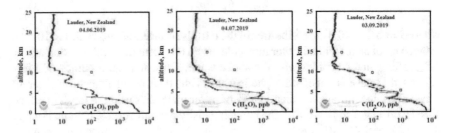

Fig. 2.5 Concentration of water molecules in the atmosphere as a result of balloon measurements at Lauder (New Zealand) [35] both during ascent and descent. Open squares correspond to formula (2.5)

two-photon absorption of the resonant vibration transition. Some results of such measurements are represented in Figs. 2.3, 2.4, and 2.5 which relate to various points of the globe. Note that balloon measurements are one of the methods of monitoring of atmospheric water [42] and give the altitude profile for atmospheric water consisting of water molecules. From this, one can obtain also the qualitative water distribution over altitudes.

Being guided by the model of the standard atmosphere, we deal with mean values of atmospheric parameters. In particular, usage of the global moisture allows one

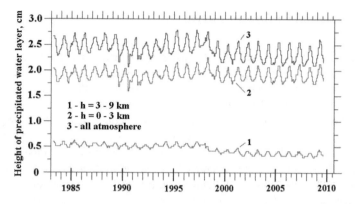

Fig. 2.6 Evolution of the average amount of atmospheric water expressed in heights of precipitated liquid water [45, 46]

Fig. 2.7 Evolution of the average moisture of atmospheric air during 1970–2010. Red—[47], green—[48], blue—[49]

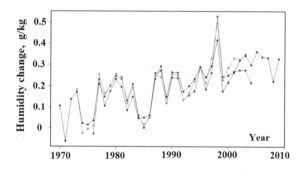

to analyze the mean distribution of the number density of water molecules over altitudes, as well as the evolution of this value in time (for example, [43, 44]). In particular, formula (2.5) gives the average altitude dependence for the number density of water molecules. From this it follows that approximately 80% of atmospheric water molecules is located at altitudes below 3 3 km. Additionally, Fig. 2.6 contains evolution in time for the water amount below and above 3 3 km that corresponds to the average parameters of formula (2.6).

We now consider the evolution of the concentration of atmospheric water in time [39]. As it follows from Fig. 2.7, the humidity of atmospheric air increases in time. The humidity change is expressed in Fig. 2.7 in units gram water per kilogram air, so that this unit of 1 g/kg corresponds to the partial water pressure 1.2 Torr. As it follows from Fig. 2.7, the rate of change of the atmospheric water amount is equal to correspondingly 0.06, 0.08, and 0.07 g/kg · per decade) according to [47–49]. These values relate to the region near the Earth's surface and give the average change of the water amount in the atmosphere (0.09 ± 0.01) Torr/decade. For comparison, the partial water pressure of the standard atmosphere is approximately 12 Torr.

2.1.3   Dynamics of Atmospheric Air

The convective motion of air is accompanied by the drift of molecules and particles of air under the action of gravitation forces and temperature gradients. We now analyze this drift of atmospheric molecules and particles. In the above consideration, the convection is created by air molecules as a mixture of nitrogen and oxygen molecules with the average mass $m = 29$ in units of atomic mass. Other components and admixtures do not contribute to the convection of air, and the mean free path of molecules is small compared with a vortex size. Therefore, components of air are captured by vortices in the course of convective motion. We divide below the air motion in the diffusion one with the mean free path of the order of a vortex size and drift motion of molecules under the action of gravitation forces. Then, distances under consideration exceed significantly the mean free path for diffusion motion, i.e. a size of vortices. In addition, the force acted on an individual molecule exceeds remarkably its weight.

Dynamics of air transport in the atmosphere [50–52] which is accompanied by convective air motion and influences various atmospheric properties. One can determine this force and the drift velocity w_L of a molecule captured by a vortex of the convective motion from the equilibrium between the diffusion and drift fluxes of molecules that are given by

$$D_L \nabla N = w_L N, \tag{2.9}$$

where N is the number density of molecules for this component, and D_L is the diffusion coefficient of air molecules due to convective motion of air. Near the Earth's surface, the diffusion coefficient D_L due to convective motion equals approximately [53, 54]

$$D_L = 4 \times 10^4 \, \text{cm}^2/\text{s} \tag{2.10}$$

One can use the diffusion character of motion of atmospheric molecules and particles captured by motion of air molecules. In particular, the equilibrium (2.9) leads to the dependence (2.3), and also to following relations for the drift velocity w_L of molecules toward the Earth and a typical time t_L of their residence in the atmosphere if they are injected in the atmosphere from the Earth's surface. We have for these parameters

$$w_L \approx \frac{D_L}{\Lambda}, \; t_L \approx \frac{\Lambda^2}{D_L}, \tag{2.11}$$

where $\Lambda = 8.4 \, \text{km}$ is the scale size according to formula (2.3). As a matter of fact, the first relation corresponds to the Einstein relation [55–57]. This description relates to convection motion of air if a typical vortex size is small compared to distances on which molecules propagate. For atmospheric air we have $w_L \approx 0.01 \, \text{cm/s}$, $t_L \approx$

200 days. This relates also to water molecules for the clear-sky weather, where the concentration of water molecules is small and interaction between them is negligible. For standard atmosphere, i.e. for an atmosphere with average parameters, where atmospheric air partakes in condensation processes, one can use formula (2.5) for the number density of atmospheric water molecules. This leads to the residence time of water molecules in a standard atmosphere

$$t_w \approx \frac{\lambda^2}{D_L} \approx 12\,\text{d} \tag{2.12}$$

This corresponds to the above value $t_w \approx (8\text{–}9)$ days, and the residence time of water molecules in the troposphere is determined by condensation processes.

From this, one can find the criterion of the convective motion for atmospheric particles. As is seen, the convective motion in the atmosphere leads to the mixing of air molecules and accompanied particles of different altitudes [58]. In the case of convective motion of air, the drift velocity of molecules w_g under the action of their weight is small compared to that resulting from their convective motion, that is

$$w_L \gg w_g, \tag{2.13}$$

One can determine the drift velocity w_g of molecules of a given type under the action of their weight on the basis of Einstein relation [55–57]

$$w_g = \frac{Dmg}{T} \tag{2.14}$$

Here, m is the mass of an admixture particle, $g \approx 980\,\text{cm/s}^2$ is the free fall acceleration, T is the air temperature expressed in energetic units, and D_m is the diffusion coefficient for a given molecule in air. This diffusion coefficient is equal within the framework of the Chapman–Enskog approximation [59–61] and the hard-sphere model for colliding atoms and molecules [61, 62]

$$D = \frac{3\sqrt{\pi T}}{8\sqrt{2\mu}N\sigma_g}, \tag{2.15}$$

where $\mu = m_1 m_2/(m_1 + m_2)$ is the reduced mass of a colliding particle and air molecule whose masses are m_1 and m_2 correspondingly, N is the number density of air molecules, σ_g is the gas-kinetic cross section for colliding atoms or molecules. In the case of air molecules, where the gas-kinetic cross section at room temperature is $\sigma_g = 38$ Å^2 [63], we have on the basis of formulas (2.14) and (2.15) at room temperature

$$N_w w_g = 6 \times 10^{12}\,\text{cm}^{-2}\text{s}^{-1}, \tag{2.16}$$

and $w_g \approx 2 \times 10^{-7}$ cm/s near the Earth's surface. As is seen, the criterion (2.13) holds true both in the troposphere and in the stratosphere.

We now consider the motion of an individual particle in convective air. Evidently, this particle is captured by air vortices if the criterion (2.13) holds true. If the opposite criterion is with respect to (2.13), its drift motion toward the Earth's surface would proceed under the action of its weight. The falling velocity w_g of a particle of a radius r with a mass density ρ of its material under the action of the particle weight is given by [64]

$$w_g = g\tau_{rel} = \frac{2\rho g r^2}{9\eta}, \tag{2.17}$$

and for water microdrops, we have $w_g/r^2 = 1.2 \times 10^6 \, s^{-1} cm^{-1}$. From this one can find that $w_L = w_g$, if $r \approx 1 \, \mu m$. From this it follows that the criterion (2.13) is violated, i.e. $w_l = w_g$ at $r = 1 \, \mu m$. Let us reduce the equilibrium altitude distribution of the number density to formula (2.5) taking it in the form [16]

$$N(h) = N_o \cdot \exp\left(-\frac{h - h_o}{\lambda_a}\right), \tag{2.18}$$

where $N_o = N(h_o)$. Then, one can represent the parameter λ_a as [16]

$$\lambda_a = \frac{\Lambda}{1 + w_g/w_L} \tag{2.19}$$

The limiting cases of this formula describe the particle fall in motionless air ($w_g \gg w_L$) and that in the case of air convective motion ($w_g \ll w_L$). From this, it follows that small particles are captured by air vortices of convective air, whereas large particles fall down under the action of their weight similar to that in the motionless air.

Thus, we have that the convective motion of air proceeds under the action of the temperature gradients in the atmosphere and gravitation forces. This motion results in air vortices, as well as air fluxes. In this process, air molecules capture molecules and particles located in air, until particle size is restricted. As a result, convection as a specific motion of air determines the motion of admixture molecules and small particles.

2.1.4 Water Microdroplet in Atmospheric Air

We now consider the behavior of an individual particle in the air for the diffusion regime of processes, where a particle size is large compared to the mean free path of air molecules in atmospheric air, that is equal to $0.1 \, \mu m$. Being guided by the motion of water particles in atmospheric air, we take into account that they are found usually in the liquid aggregate state. Thus, we consider below the behavior of water microdroplets in air. Note that a typical size of water microdroplets which form clouds lies within the limits $(1–10)\,\mu$m. For such microdroplets, we have for the Reynolds number which describes the motion of atmospheric air near an individual microdroplet

$$\mathrm{Re} = \frac{v_l r}{\nu} \sim 0.01 \tag{2.20}$$

We thus deal with a laminar motion of air near the microdroplet.

Let us analyze the resisting action from a microdroplet which is located in moving air. Taking $\mathbf{v}_o$ as the air velocity and $\mathbf{v}$ as the microdroplet motion, we have the motion equation for the microdroplet as

$$M\frac{d\mathbf{v}}{dt} = \mathbf{F} = 4\pi r\eta(\mathbf{v} - \mathbf{v}_o) = -\frac{M(\mathbf{v} - \mathbf{v}_o)}{\tau_{\mathrm{rel}}},$$

where F is the Stokes force [64], M is the mass of the microdroplet, and τ_{rel} is its relaxation time which according to [65] is given by

$$\tau_{\mathrm{rel}} = \frac{2\rho r^2}{9\eta} \tag{2.21}$$

Here, $\rho = 1\,\mathrm{g/cm^3}$ is the mass water density, and $\eta = 1.85\,\mathrm{g/(cm\,s)}$ is the air viscosity coefficient at room temperature. If a microdroplet consists of liquid water, we have from this formula $\tau_{\mathrm{rel}}/r^2 = 1.2 \times 10^3\,\mathrm{s/cm^2}$. For parameters of the model of standard atmosphere and a typical size $r = 8\,\mu$m of liquid microdrops in clouds, we have $\tau_{\mathrm{rel}} \approx 7 \times 10^{-4}$ s. In the case of droplet falling, one can determine the velocity of droplet fall w_g from equality of the weight $P = Mg$ (M is the droplet mass, and $g = 980\,\mathrm{cm/s^2}$ is the free fall acceleration) and the Stokes formula which characterizes the relaxation force for the droplet and gives Mw_g/τ_{rel}. This equality gives [64]

$$w_g = \frac{2\,g\rho r^2}{9\eta} \tag{2.22}$$

Let us compare this relaxation time with a typical time τ_l of velocity variation in vortices of a size l which is estimated as $\tau_l \sim l/v_l$. Restricting ourselves by large vortices of size L which determine the air transport due to convection, we have $\tau_L \sim L/v_L \sim 0.06$ s for atmospheric air near the Earth's surface. Requiring $\tau_{\mathrm{rel}} \ll \tau_L$, we have that fast establishment of equilibrium for water microdrops in atmospheric air according to this criterion takes place at sizes $r \ll 70\,\mu$m. The range of microdrop sizes $r \sim (1–10)\,\mu$m is of interest now; such microdroplets are captured by air vortices in the troposphere.

Let us consider other relaxation processes related to the thermal regime of the behavior of a water microdroplet in atmospheric air. If the temperature T_d of a water microdroplet differs from the temperature T_o of the surrounding air, the heat flux q arises

$$q = -\kappa \frac{dT}{dR}$$

that equalizes the air and microdroplet temperatures. Here, κ is the air thermal conductivity, and R is a distance from the droplet center. The power Q which is transferred from air to the drop or vice versa is equal to

$$P = 4\pi R^2 \kappa \frac{dT}{dR} \tag{2.23}$$

Solving equation $Q = \text{const}$ with the boundary conditions $T(r) = T_d$, $T(\infty) = T_o$, we have for the current temperature $T(R)$

$$T(R) = T_o + \frac{T_d - T_o}{R} \tag{2.24}$$

In considering the case where this distribution is established fast, one can obtain for the heat power from a heated microdroplet to the surrounding air on the basis of formula (2.23)

$$P = 4\pi r \kappa \Delta T, \quad \Delta T = |T_d - T_o|, \tag{2.25}$$

where r is the droplet radius.

We have finally, the temperature T_d tends to the temperature T_o of the surrounding air. This process is described by equation

$$C_p \frac{d\Delta T}{dt} = -P = -4\pi r \kappa \Delta T, \tag{2.26}$$

where the droplet heat capacity is

$$C_p = c_p \cdot \frac{4\pi r^3 \rho}{3}$$

Here, the specific heat capacity for liquid water is $c_p = 1 cal/(g \cdot K)$. Represent equation (2.26) as

$$C_p \frac{d\Delta T}{dt} = -P = -4\pi r \kappa \frac{\Delta T}{\tau_r}, \quad \tau_r = \frac{r^2 \rho c_p}{3\kappa}, \tag{2.27}$$

where τ_r/r^2 is the relaxation time for heat. In particular, for water microdroplets in atmospheric air, we have $\tau_r/r^2 = 5.5 \times 10^3$ s/cm^2. For a size of a water microdroplet $r = 8\,\mu\text{m}$ which corresponds to clouds, we have $\tau_r = 0.9\,\text{ms}$. This relaxation time is also the evaporation time of this drop in atmospheric water which does not contain water molecules far from the droplet.

It is essential that the processes of molecule attachment to microdroplets and evaporation of molecules are accompanied by heat processes. This may lead to a change in the character of these processes. Let us analyze these processes near a microdrop of a radius r, taking c_o as the concentration of water molecules, and c_* as the concentration of water molecules at the saturated vapor pressure for a given temperature. For a non-saturated vapor $c_o < c_*$, a microdrop evaporates. For the diffusion regime of this process, the molecule flux i from the microdroplet surface is equal at a distance R from the particle surface

$$i = -D_w N_w \frac{dc}{dR}, \tag{2.28}$$

where D_w is the diffusion coefficient for a water molecule in atmospheric air, which is equal under normal conditions $D = 0.22 \, \text{cm}^2/\text{s}$ [63, 66], and N_w is the number density of water molecules. A number of water molecules which intersect a sphere of a radius R per unit time is

$$j = 4\pi R^2 \cdot i$$

One can consider this relation as an equation for the concentration $c(R)$ of water molecules because the flux j is determined by the processes of molecule attachment and evaporation from the microdroplet and therefore, it is independent of R. Solving this equation, one can obtain [67]

$$c(R) = c_o + \frac{r \Delta c}{R}, \quad j = 4\pi D_w N_w r \Delta c, \tag{2.29}$$

where $\Delta c = c_* - c_o$.

This corresponds to the diffusion regime of attachment of atoms or molecules to a droplet. In this regime, we have for the flux J_{at} of molecule attachment to the droplet surface

$$J_{at} = 4\pi D_w N_w r, \tag{2.30}$$

where D_w is the diffusion coefficient of attachment molecules in air, N_w is the number density of free water molecules far from the droplet, and r is a droplet radius. This formula is known as the Smoluchowski formula [68]. This formula allows one to find the rate of molecule evaporation from the droplet surface. Indeed, the equilibrium is an establishment for a droplet and free molecules in a saturated vapor, where the number density of molecules is equal to N_{sat}, that is the saturated vapor pressure. This means that the number of molecules evaporated from the droplet per unit time is equal to those attached to the droplet surface. Thus, we have for the rate of droplet evaporation

$$J_{ev} = 4\pi D_w N_{sat} r \tag{2.31}$$

Correspondingly, the balance equation for a droplet consisting of n bound molecules has the form

$$\frac{dn}{dt} = J(n), \; J(n) = 4\pi D_w r(N_w - N_{\text{sat}}), \tag{2.32}$$

and $J(n)$ is the total rate of the change of the number of molecules for a droplet consisting of n molecules with accounting for processes of droplet evaporation and molecule attachment. We remember that these formulas relate to the diffusion regime of processes where a droplet radius r_o is large compared to the mean free path of molecules λ in atmospheric air.

These processes of droplet evaporation and molecule attachment are accompanied by the processes of absorption and release processes, and the energy change in one act, i.e. the molecule binding energy ε_o exceeds remarkably a typical thermal energy T of molecules ($\varepsilon_o \gg T$). Hence, a thermal effect may be essential, and we analyze it below. Let us introduce the difference Δ between the droplet temperature and the gaseous one far from the droplet in accordance with formula (2.25). This creates the heat flux P given by formula (2.26) that is compensated by the energy flux $\varepsilon_o J_{at}$ due to processes of droplet evaporation and molecule attachment. From the equality of these fluxes $P = \varepsilon_o J_{at}$, one can find the temperature difference.

From this, one can determine the temperature change for a water droplet as a result of molecule attachment to the droplet surface. Indeed, due to a thermal equilibrium, the released power as a result of molecule attachment to the droplet surface $\varepsilon_o J_{at}$ is equal to the released thermal energy P given by formula (2.26), where J_{at} is determined by the Smoluchowski formula (2.30). As a result, we have for the temperature difference of a droplet and a gas far from the droplet

$$\Delta T = \frac{P}{4\pi r \kappa} = \frac{\varepsilon_o J_{at}}{4\pi r \kappa} = \frac{\varepsilon_o D_w (N_w - N_{\text{sat}})}{\kappa} \tag{2.33}$$

One can see that the transition to a supersaturated vapor leads to the change of the temperature difference.

Because of a large binding energy ε_o of a bound molecule of the droplet compared to a typical thermal energy T of free water molecules, the feedback may be remarkable for the thermal effect under consideration. Taking into account that the air temperature near the droplet differs by ΔT from that far from the droplet, we have in the case $\Delta T \ll T$ instead of the balance equation (2.33)

$$\kappa \Delta T = \varepsilon_o D_w (N_w - N_{\text{sat}}) \exp\left(-\frac{\varepsilon_o \Delta T}{T^2}\right). \tag{2.34}$$

As a result, we have instead of formula (2.33)

$$\Delta T = \frac{\varepsilon_o D_w (N_w - N_{\text{sat}})}{\kappa \Phi(T)}, \; \Phi(T) = 1 + \left(\frac{\varepsilon_o}{T}\right)^2 \frac{\varepsilon_o D_w |N_w - N_{\text{sat}}|}{\kappa} \tag{2.35}$$

Fig. 2.8 Thermal factor $\Phi_o(T)$ that accounts for heating of the water droplet in atmospheric air as a result of the growth process [69, 70]

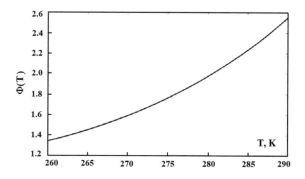

One can see that the feedback may be essential only far from the equilibrium between the evaporation and attachment processes. In particular, if a droplet is located in a vacuum, formula (2.35) takes the form

$$\Phi_o(T) = 1 + \left(\frac{\varepsilon_o}{T}\right)^2 \frac{\varepsilon_o D_w | N_{\text{sat}}}{\kappa} \tag{2.36}$$

In particular, taking $\varepsilon_o = 0.43$ eV [71], one can obtain $\Phi_o(T) = 1.7$ at the temperature $T = 273$ K and atmospheric pressure, where $D_w = 0.22 \,\text{cm}^2/\text{s}$ [63, 66], $N_{\text{sat}} = 1.6 \times 10^{17} \,\text{cm}^{-3}$, and $\kappa = 2.4 \times 10^{-4}$ W/(cm K). Figure 2.8 gives the dependence $\Phi_o(T)$.

As is seen, evaporation and growth of water microdroplets are accompanied by processes of the energy absorption and release. The power of the total process is equal to $Q = \varepsilon_o j$, where j is the flux of molecules, and $\varepsilon_o = 0.43$ eV [71] is the binding energy per water molecule. From this, we have in accordance with formula (2.24) for the temperature change

$$\frac{\Delta T}{\Delta c} = -\frac{D_w N_a \varepsilon_o}{\kappa}, \tag{2.37}$$

irrespective of the relation between the number density of free molecules N_w and N_{sat}, that corresponds to the saturated vapor pressure. This formula is conserved also in the case where free molecules are reflected partially from the droplet surface.

We also consider the case where heat release from a droplet results from its radiation as a blackbody. Let us compare the heat flux due to thermal conductivity P and that due to radiation of the microdrop J_r. Assuming a drop radius r to be large compared to the wavelength of radiation and considering and water surface as a blackbody, we have for exchange by radiation between the microdroplet and surrounding air that is optically thick for radiation. Then we have

$$J_r = 4\sigma T^3 \Delta T \cdot 4\pi r^2,$$

where σ is the Stefan–Boltzmann constant, $4\pi r^2$ is the area of the droplet surface which is a blackbody. In this case, the ratio of heat release from the droplet due to thermal radiation and thermal conductivity of air is equal to

$$\xi = \frac{J_r}{P} = \frac{4\sigma T^3 r}{\kappa} \tag{2.38}$$

In particular, for the droplet size of a cumulus cloud $r = 8\,\mu m$ at atmospheric pressure and the temperature $T = 273\,K$ we have $\xi = 1 \times 10^{-3}$. We also consider the case of an optically thick air layer, where under the above conditions cooling of a droplet equals

$$\Delta T = \frac{\sigma T^4 \cdot r}{\kappa} \tag{2.39}$$

For a droplet of a cumulus cloud at the temperature $T = 273\,K$ this formula gives $\Delta T = 0.1\,K$. In addition, this gives near the Earth's surface ($T = 288\,K$, $\kappa = 2.5 \times 10^{-4}\,W/(cm\ K)$) $\Delta T = 0.15\,K$. One can see that the droplet emission is a slow process. Thus, from this analysis one can conclude that the radiative transfer is a weak process for water microdroplets in atmospheric air.

2.2 Global Earth's Energetics

2.2.1 Contemporary Evolution of Global Temperature

The global temperature is introduced as the Earth's surface temperature averaged over geographic coordinates, time of day, and season. The global temperature is the characteristics of the thermal state of the Earth's surface, such that its change in time describes the evolution of the thermal state of our planet. The evolution of the global temperature is accompanied by large fluctuations because of large variations of the temperature for a given geographical point in summer and winter, as well as a large difference of this parameter for different geographical points at the same time. As a result, typical fluctuations of the global temperature are of the order of several °C, whereas variations of the global temperature inside one century are measured in 0.1 °C. But the object of our consideration is the change of the global temperature whose fluctuations are equal of the order (0.1–0.2) K. The method has been developed by NASA [72] for the determination of the change of the global temperature ΔT by comparison of the temperature difference for a given point on the Earth's surface in different years, but at the same time of day and season. The subsequent averaging of this value over the geographic coordinate and time allows us to follow the evolution of the global temperature with fluctuations of the order (0.1–0.2) K.

This method was developed within the framework of the NASA program (Goddard Institute for Space Studies—GISS) [73–78]. The change of the global temperature,

Fig. 2.9 Evolution of global temperature averaging over 1 and 5 years [73] (**a**), and over 5 and 15 years [74, 75] (**b**). The arrows indicate the time period from 1951 to 1980, during which the global temperature is considered constant and which can be used as a reference point

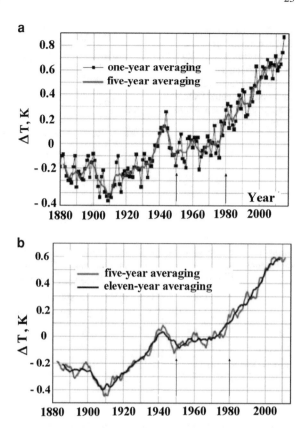

which is performed on the basis of the above algorithm, is represented in Fig. 2.9. This figure is constructed on the basis of temperature measurements during the last 150 years for different parts of the globe. For this goal, information was used from more than six thousand meteorological stations that existed in the late of the nineteenth century, and also data of last time from satellites. As can be seen, the global temperature has changed non-monotonically during the period under consideration, and this change is relatively small in the first stage of the period, whereas the global temperature increases sharply in the last about 40 years. This part of temperature evolution is given separately in Fig. 2.10 [17]. Treatment of data of Fig. 2.9 for significant Earth's warming gives the following rate of variation of the global temperature [17]

$$\frac{d\Delta T}{dt} = (18 \pm 3) \times 10^{-3}\,\text{K/year},\qquad(2.40)$$

as it follows from the processing of these data.

Using the average global parameters of the Earth's surface, it should be noted that they really depend on the geographical location of the area under consideration. Similarly, the change in the global temperature occurs uniformly across the globe.

Fig. 2.10 Evolution of global temperature in the period of strong warming [17] constructed on the basis of [75]. The temperature variation is counted from the average global temperature during 1950–1980

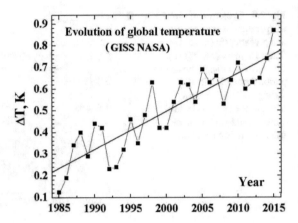

Table 2.1 The change in the global temperature of the Earth expressed in °C, where the average temperature for the twentieth century is taken as zero [79, 80]

	Total globe May 2018	Northern Hemisphere May 2018	Southern Hemisphere May 2018	Total globe May 2019	Northern Hemisphere May 2019	Southern Hemisphere May 2019
Land	1.21	1.27	1.06	1.16	1.25	1.13
Oceans	0.60	0.60	0.54	0.73	0.81	0.69
Land+Oceans	0.77	0.91	0.62	0.85	0.93	0.77

Table 2.1 contains the average change of the global temperature during May 2018 and 2019 for land and oceans separately, as well as in the Northern and Southern hemispheres. As it follows from this Table, the greatest changes relate to the land of the Northern hemisphere, where the main industrial potential is concentrated. However, these changes cannot be directly linked to the energy release as a result of human industrial and agricultural activity, because this energy release is relatively small. Evidently, human activity changes the conditions on the Earth's surface and in the atmosphere, which affect the energy processes.

2.2.2 Earth's Temperature in the Past

A certain understanding of the nature of the evolution of the Earth's temperature follows from the isotopic analysis of sediments, which are taken from the Earth's interior. Such investigations are connected with geo-paleontology [81] and allow one to determine the local temperature at a certain time. We below consider an example associated with the change in the local temperature of Antarctica in the past. In this case, air bubbles extracted from ice deposits of the meteorological station "Vostok" [82, 83] were examined, and because the depth of a certain deposit is connected with

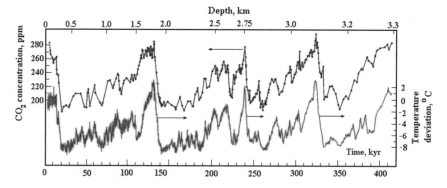

Fig. 2.11 Evolution of the Earth's temperature in the past in the area of the meteorological station "Vostok" (Antarctica), as well as the concentration of carbon dioxide, which is obtained [82, 83] from the analysis of bubbles inside ice pieces extracted from a corresponding depth

a time of bubble formation, it gives the concentration of atmospheric carbon dioxide at that time. The air temperature follows from the analysis of the oxygen isotope ^{18}O which is distributed among oxygen and carbon dioxide molecules depending on the air temperature. The results of this analysis are given in Fig. 2.11.

In considering the Earth's temperature in the past, note that in this case it is possible to determine the local temperature only. As it follows from Fig. 2.11, the maximum amplitude of temperature variation reaches 14 K. Probably, large fluctuations of the Earth's temperature are connected with glacial epochs. The period of a glacial epoch is of the order of 100 kyr and the nature of existence of the glacial epochs is explained according to the Milankovitch theory [84, 85] due to the character of Earth's motion in its rotation around the Sun along the elliptic orbit. This orbit is close to a ring, and the eccentricity of the Earth's orbit that characterizes the difference of this orbit and a ring one varies between 0.005 and 0.028, at the contemporary value of 0.017. The period of oscillations due to variation of the eccentricity is approximately 100 000 000 year that corresponds to the glacial period.

Evidently, oscillations of parameters of the Earth's orbit around the Sun are determined by the interaction with the gravitational fields of Jupiter and Saturn and a mixture of these oscillations leads to the irregular variation of the global temperature in time [86–88]. Thus, in this case variations of the solar irradiance for the Earth's surface result from variations of the Sun-Earth distance [89], and this is the reason for oscillations of the global temperature.

Note that various mechanisms are responsible for the variation of global and local temperatures of the Earth for some time periods. In the above case, the change of the global temperature follows from the variation of the distance from the Sun. From Fig. 2.9, it follows that variations in the greenhouse effect lasting several decades have another nature, and the time dependence of the global temperature in Fig. 2.9 becomes smooth after averaging over several years. We give in Fig. 2.12 variations of the local temperature near the Greenland during the last millenniums [90]. This

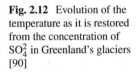

Fig. 2.12 Evolution of the temperature as it is restored from the concentration of SO_4^2 in Greenland's glaciers [90]

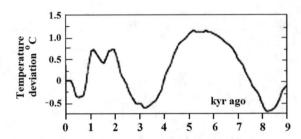

temperature is determined on the basis of the mass of sulfate salts for the various layers of the Greenland glacier and, probably, was governed by explosions of volcanoes. In this case, sub micron particles from volcanoes penetrate the stratosphere or upper atmosphere and, are located there long enough, influencing the Earth's albedo.

Variations of the Earth's temperature in the last millennium may be connected to a great extent with variations of the solar irradiance [91]. In particular, a warm period was observed during 11–14 centuries in England [92], when the grape was ripened and the Vikings occupied Greenland. During the next cold period in the fourteenth century, the river Thames was frozen each year and the Vikings left Greenland [92]. Variations in solar irradiation are assumed to be a reason of so-called Maunder minimum which lasted from 1645 to 1715 [93]. One can connect the solar activity with some processes inside the Sun [94, 95], that, in turn, influences the Earth's temperature. In particular, the amplitude of the temperature change during the Maunder minimum is approximately 1 K [91] that corresponds to the variation of the global temperature in the last century. One can see that the above facts are confirmed by the temperature evolution given in Fig. 2.12 for indicated times.

We consider above the mechanisms of the global temperature change as a result of the change of the solar irradiance both due to processes inside the Sun and variations of the distance between the Earth and Sun. Among other mechanisms, we mark the action of particles of meteoric haze [96] which absorb and scatter solar radiation that leads to a decrease of the global temperature. Let us estimate an amount of the meteoric dust which is able to decrease the solar flux. If we collect this dust in one layer, its thickness would be of the order of the radiation wavelength. One can see that the density on the way of photons is equal in this case of the order of $(10^{-5}$ to $10^{-4})\,g/cm^2$, that exceeds remarkably its density in an interplanetary space of the order of $(10^{-18}$ to $10^{-17})\,g/cm^2$ and is able to influence the solar flux to the Earth's atmosphere. In particular, due to the generation of dust, the Halley comet creates periodic variations of the global temperature of amplitude 0.08 °C with a period of (72 ± 5) year [97]. In addition, penetrating the Earth's atmosphere, cosmic dust particles may be condensation nuclei in formation of water microdroplets [98, 99].

Along with particles of cosmic dust in an interplanetary space, cosmic rays of galactic and solar origin may influence the atmospheric phenomena which determine the thermal balance of our planet. As it is discussed from 1959 [100], atmospheric ions formed under the action of cosmic rays become nuclei of condensation and in

Fig. 2.13 Correlation between variations in the flux of cosmic rays and the change of cloudiness in the lower troposphere [92, 106, 107]

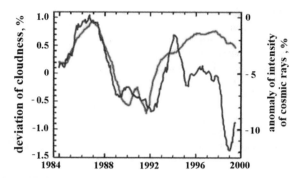

this manner they are responsible for cloud formation. If this statement holds true, the cloudiness must be correlated with the intensity of cosmic rays. Anomalies in the intensity of cosmic radiation modulated by the solar plasma and those for the atmospheric coverage with clouds are compared in Fig. 2.13. The active analysis of this problem was made by Svendsmark et al. [101–104] that allows one to understand the influence of cosmic radiation on the Earth's climate in the past and also was a basis of an experiment with synchrotron radiation from the CERN accelerator [105].

One can see from Fig. 2.13 that the correlation between the cloudiness and the intensity of cosmic rays took place in the 22 cycles of solar activity during 1981–1992, whereas this correlation is absent in the subsequent 23rd solar cycle. Explanation of this fact is that condensation during the 23rd solar cycle results due to radicals resulting from human activity, whereas their amount during the 22nd solar cycle was not enough to dominate. This means that the atmospheric pollution in our time attains the level where natural conditions become not enough for condensation processes. It should be added that the analysis of condensation processes due to cosmic rays causes discussions and objections (for example, [108–113]). In spite of these contradictions, the analysis of the role of cosmic rays in cloud processes allowed one to understand deeply the character of atmospheric energetic processes and to extract the sensitive elements in the chain of these processes.

2.2.3 Carbon Dioxide and Earth's Temperature

As it follows from Fig. 2.11, the correlation takes place between the concentration of the atmospheric carbon dioxide and the global temperature. The characteristic of this connection is ECS (Equilibrium Climate Sensitivity) [114] that is the change of the global temperature as a result of doubling of the concentration of atmospheric carbon dioxide. It should be noted that considering the ECS, we take the concentration of CO_2 molecules as the parameter of the thermal Earth's state, rather than this is the reason of an increasing global temperature, an increase of the atmospheric CO_2 amount. Thus, the Equilibrium Climate Sensitivity is introduced as [114]

Fig. 2.14 Concentration of CO_2 molecules in atmospheric air during the last half-century (**a**) and for the last 5 years (**b**) according to [115, 116]; open circles correspond to an average during a month, and filled squares relate to averaged data for a year (one-half year before and after indicated data)

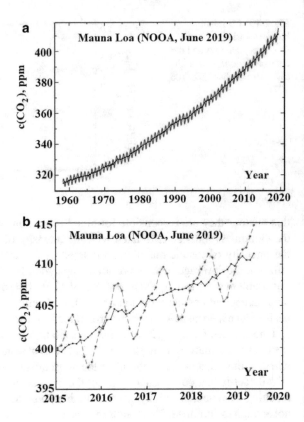

$$ECS \equiv \ln 2 \frac{d\Delta T}{d \ln c(CO_2)}, \tag{2.41}$$

where ΔT is a change of the global temperature, and $c(CO_2)$ is the concentration of molecules of atmospheric carbon dioxide.

In this consideration, we use that the residence time of a carbon dioxide molecule in the atmosphere is about 5 years. This means that CO_2 molecules are mixed with atmospheric air completely, i.e. the concentration of atmospheric CO_2 molecules is identical for any geographical points which are located far from sources and absorbers of carbon dioxide. We will use for this goal the results of the Mauna Loa observatory (Hawaii, USA) [117–122] as the best source for monitoring of atmospheric CO_2 molecules. This observatory is located at altitude 3400 m above the sea level with globe coordinates ($19°32'N$, $155°35'W$) and far from sources or absorbers of carbon dioxide [123]. The measuring equipment is found in four towers of the height of 7 m and one tower of the height of 27 m, so that the influence of the Earth's surface on measurements is weak.

Some results of this monitoring are given in Fig. 2.14 [115, 116]. According to these measurements, the concentration of carbon dioxide molecules in atmospheric air increases from 316 ppm in 1959 up to 412 ppm in 2020. In addition, the rate of an increase in the carbon dioxide concentration grows in time from 0.7 ppm/year in 1959 up to approximately 2.1 ppm in 2020. Note also that in 1750, the concentration of CO_2 molecules was equal to (277 ± 3) ppm, and in 1870 this value was (288 ± 3) ppm [124]. The data of Fig. 2.14 allows us to determine the ECS on the basis of NASA data for monitoring of the global temperature ΔT and the concentration of atmospheric CO_2 molecules. In particular, for the last 5 years from 2015 up to 2020 according to Fig. 2.13b, the concentration of atmospheric CO_2 molecules has varied from 400 ppm up to 412 ppm, so that

$$\frac{d \ln c(CO_2)}{dt} = 0.0006 \, \text{year}^{-1} \tag{2.42}$$

From this, it follows that the doubling of the concentration of atmospheric CO_2 molecules proceeds through $(110\text{--}120)$ year.

In addition, from formulas (2.40) and (2.41), it follows that

$$\text{ECS} \equiv \ln 2 \frac{d\Delta T}{d \ln c(CO_2)} = (2.1 \pm 0.4) \, \text{K} \tag{2.43}$$

Note that this value results from NASA monitoring for the global temperature and the concentration of atmospheric CO_2 molecules, rather than from some models.

2.2.4 Energetic Balance for the Earth and Its Atmosphere

The energetic balance of the Earth and its atmosphere includes three types of processes, namely, radiative processes involving solar radiation and its secondary processes, processes of infrared radiation for the Earth and its atmosphere, and heat transport from the Earth to the atmosphere. Figure 2.15a contains the average powers of various channels for energetic processes which involve the Earth and atmosphere as a whole. These data are obtained in NASA measurements [125] and are taken from books [16, 126, 127]. It is convenient to convert the global powers into the average energy fluxes though these energy fluxes are different for different points of the globe. We give in Fig. 2.15b the average fluxes for energetic processes with participation of the Earth and atmosphere.

The nearby results are obtained by World Meteorological Organization (WMO) [128, 129] and just these data are used overall including monographs [12, 13, 130–140] and are analyzed in papers [141–145]. The above data of NASA and WMO and their comparison allow one to make some conclusions. Below we consider this briefly. First, as it follows from diagrams of Fig. 2.15, the power of infrared radiation absorbed by the Earth is twice compared to that of solar radiation in the visible spectrum range that testifies about the importance of infrared radiation in the Earth's energetic balance. Second, the Earth emits almost like a blackbody. Indeed, let us

Fig. 2.15 Powers of some
energetic processes which
involve the Earth and
atmosphere as a whole and
are expressed in 10^{16} W (**a**),
and average energy fluxes in
W/m² for these channels
each of which is the ratio of
the above power to the
Earth's surface
$S = 5.1 \times 10^{14}$ m² (**b**)

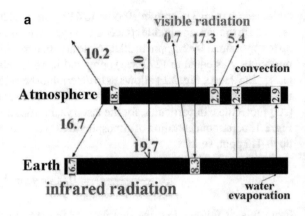

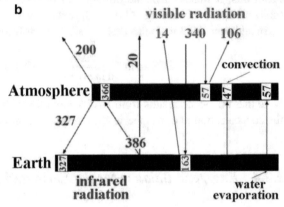

use the Stefan–Boltzmann law for the radiative flux J_E from the Earth's surface in
the infrared spectral range in the form

$$J_E = \sigma T_E^4, \tag{2.44}$$

where $\sigma = 5.67 \times 10^8$ W/m² is the Stefan–Boltzmann constant, and T_E is the global
temperature. Taking the radiative flux $J_E = 386$ W/m² according to Fig. 2.15, one
can obtain the Earth's temperature $T_E = 287$ K, whereas according to the model of
standard atmosphere this value is $T_E = 288$ K. If we exclude the atmosphere as a
source of infrared radiation from its energetic balance accounting only its interaction
with solar radiation, one can obtain for the Earth's temperature $T_E = 232$ K. It should
be added to this that if the atmosphere does not act on indicated solar radiation, all
the solar radiation is absorbed by the Earth and it emits like a blackbody, the global
temperature would be $T = 278$ K. All this shows the atmosphere role in the Earth's
energetic balance.

Fig. 2.16 Channels of the global energy resulting from human activity annually and expressed in 10^{18} J [146, 147]

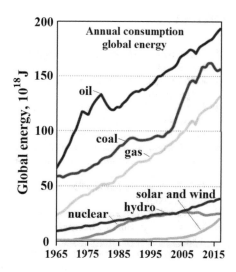

One can see from diagrams in Fig. 2.15 that the infrared radiation of the atmosphere outside is separated from that toward the Earth, and only approximately 5% of the radiative flux emitted by the Earth passes outside. Hence, one can conclude that the optical thickness of the atmosphere in the infrared spectrum range is large. Therefore, in considering atmospheric emission toward the Earth, one can ignore processes of heat transport inside the atmosphere, as well as for its outgoing emission. In this context, it is important that atmospheric emission is created mostly by atmospheric water, both in the form of free water molecules and in the form of microparticles or microdroplets which constitute clouds. All these facts testify about the importance of the atmosphere as a source of infrared radiation in the energetic balance of the Earth, as well as the role of atmospheric water molecules and microdroplets as elementary radiators of the atmosphere.

It is of interest to compare the energetic balance of the Earth with that as a result of human activity. The contemporary released power as a result of human activity is about 2×10^{13} W or to the average energy flux of 0.04 W/m^2 that is approximately four orders of magnitude below the solar energy flux penetrating in the Earth's atmosphere. Figure 2.16 contains the main channels of energy consumption as a result of industrial human activity. As is seen, the main contribution to global human energetics follows from combustion of fossil fuels (petroleum, coal, and natural gas). This channel includes approximately 80% of consumption energy and this energy grows by 2.3% per year. Though the power of renewable sources of energy grows sharper with the rate below 20% per year, the energy of fossil fuels will be the basic source of human energetics during the twenty-first century.

References

1. J.T. Houghton, *The Physics of Atmospheres* (Cambridge University Press, Cambridge, 1977)
2. J.V. Iribarne, H.P. Cho, *Atmospheric Physics* (Reidel Publishing, Dordrecht, 1980)
3. R.G. Fleagle, J.A. Businger, *Introduction to Atmospheric Physics* (Academic Press, San Diego, 1980)
4. R.M. Goody, Y.L. Yung, *Principles of Atmospheric Physics and Chemistry* (Oxford University Press, New York, 1995)
5. M.L. Salby, *Fundamentals of Atmospheric Physics* (Academic Press, San Diego, 1996)
6. J.H. Seinfeld, S.N. Pandis, *Atmospheric Chemistry and Physics* (Wiley, New York, 1998)
7. D.G. Andrews, *An Introduction to Atmospheric Physics* (Cambridge University Press, Cambridge, 1999)
8. D.G. Andrews, *An Introduction to Atmospheric Physics* (Cambridge University Press, Cambridge, 2000)
9. J.H. Seinfeld, S.N. Pandis, *Atmospheric Chemistry and Physics* (Wiley, Hoboken, 2006)
10. J.M. Walace, R. Hobbs, *Atmospheric Science. An Introductory Survey.* (Elsevier, Amsterdam, 2006)
11. M.H.P. Ambaum, *Thermal Physics of the Atmosphere* (Wiley-Blackwell, Oxford, 2010)
12. M.L. Salby, *Physics of the Atmosphere and Climate* (Cambridge University Press, Cambridge, 2012)
13. I. Lagzi (EA), *Atmospheric Chemistry.* (Institute of Geography and Earth Science, Budapest, 2013)
14. R. Caballero, *Physics of the Atmosphere* (IOP Publishing, Bristol, 2014)
15. G. Visconti, *Fundamentals of Physics and Chemistry of the Atmosphere* (Springer Nature, Switzerland, 2017)
16. B.M. Smirnov, *Microphysics of Atmospheric Phenomena* (Springer Atmospheric Series, Switzerland, 2017)
17. B.M. Smirnov, *Physics of Global Atmosphere.* (Intellect, Dolgoprudny, 2017; in Russian)
18. https://en.wikipedia.org/wiki/Troposphere
19. *U.S. Standard Atmosphere.* (Washington, U.S. Government Printing Office, 1976)
20. P.M. Morse, *Thermal Physics* (Benjamin Inc., New York, 1964)
21. D. Haar, H. Wergeland, *Elements of Thermodynamics* (Addison-Wesley, Reading, 1967)
22. R. Kubo, *Thermodynamics* (North Holland, Amsterdam, 1968)
23. C. Kittel, H. Kroemer, *Thermal Physics* (Wiley, New York, 1980)
24. I.A. Shiklomanov, in *Water in Crisis: A Guide to the World's Fresh Water Resources*, ed. by P.H. Gleick (Oxford University Press, Oxford, 1993), pp. 13–24
25. I.A. Shiklomanov, J.C. Rodda (eds.), *World Water Resources at the Beginning of the Twenty-First Century* (Cambridge University Press, Cambridge, 2003)
26. R.W. Healy, T.C. Winter, J.W. Labaugh, O.L. Franke, *Water Budgets: Foundations for Effective Water-resources and Environmental Management* (U.S. Geological Survey Circular 1308, Reston, Virginia, 2007)
27. http://en.wikipedia.org/wiki/Atmosphere-of-Earth
28. A. Baumgartner, E. Reichel, *The World Water Balance* (Elsevier, Amsterdam, 1975)
29. J.P. Peixoto, A.H. Oort, *Physics of Climate* (American Institute of Physics, Washington, 1992)
30. K.E. Trenberth, L. Smith, T. Qian, et al., J. Hydrometeorol. **8**, 758 (2007)
31. http://en.wikipedia.org/wiki/water-circle
32. http://water.usgs.gov/edu/watercyrcleatmosphere.html
33. https://en.wikipedia.org/wiki/Moisture
34. https://en.wikipedia.org/wiki/Dew-point
35. https://www.esrl.noaa.gov/gmd/dv/iadv/graph.php?code=HIH.program=wvap.type=vp
36. https://tamino.wordpress.com/2010/08/08/urban-wet-island/
37. http://www.c3headlines.com/greehouse-gases-atmosphereco2methanewater-vapor
38. http://www.c3headlines.com/natural-negativepositive-feedback

39. https://tamino.wordpress.com/2011/05/17/hot-and-wet/
40. https://en.wikipedia.org/wiki/Weather-balloon
41. https://public.wmo.int/en/resources/bulletin/observing-water-vapour
42. *Monitoring Atmospheric Water Vapours*, ed. by N.K Ampfer (Springer Nature, Bern, 2013)
43. R.F. Alder et.al., J. Hydrometeor. **4**, 147 (2003)
44. G.J. Huffman, R.F. Alser, D.T. Bolvin, G. Gu, Geoph. Res. Lett. **36**, L17808 (2009)
45. http://www.climate4you.com/GreenhouseGasses.htm
46. http://atmospheres.gsfc.nasa.gov/meso/index.php
47. A. Dai, J. Clim. **19**(3589), 963606 (2006)
48. K.M. Willett, P.D. Jones, N.P. Gillett, P.W. Thorne, *J. Clim.* **21**, 5364 (2008)
49. D.I. Berry, E.C. Kent, *Bull. Am. Meteor. Soc.* **90**, 645 (2009)
50. E. Lorenz, *The Nature and Theory of General Circulation of the Atmosphere* (World Metheorological Organization, Geneva, 1967)
51. G.T. Csanady, *Turbuletnt Diffusion in Environment* (Holland, Reidel, Dordrecht, 1973)
52. D.J. Tritton, *Physical Fluid Dynamics* (Claredon Press, Oxford, 1988)
53. T. Shimazaki, A.R. Laird, Radio Sci. **7**, 23 (1972)
54. M.J. McEwan, L.N. Phillips, *Chemistry of the Atmosphere* (Edward Arnold, London, 1975)
55. A. Einstein, Ann. Phys. **17**, 549 (1905)
56. A. Einstein, Ann. Phys. **19**, 371 (1906)
57. A. Einstein, Zs.für Electrochem. **14**, 235 (1908)
58. https://en.wikipedia.org/wiki/Atmospheric-circulation
59. S. Chapman, T.G. Cowling, *The Mathematical Theory of Non-uniform Gases* (Cambridge University Press, Cambridge, 1952)
60. J.H. Ferziger, H.G. Kaper, *Mathematical Theory of Transport Processes in Gases* (North Holland, Amsterdam, 1972)
61. M. Capitelli, D. Bruno, A .Laricchiuta, *Fundamental Aspects of Plasma Chemical Physics. Transport* (Springer, New York, 2013)
62. B.M. Smirnov, *Physics of Ionized Gases* (Wiley, New York, 2001)
63. B.M. Smirnov, *Reference Data on Atomic Physics and Atomic Processes* (Springer, Heidelberg, 2008)
64. L.D. Landau, E.M. Lifshits, *Fluid Mechanics* (Pergamon Press, London, 1959)
65. B.M. Smirnov, *Nanoclusters and Microparticles in Gases and Vapors* (DeGruyter, Berlin, 2012)
66. N.B. Vargaftic, *Tables of Thermophysical Properties of Liquids and Gases* (Halsted Press, New York, 1975)
67. B.M. Smirnov, *Clusters and Small Particles in Gases and Plasmas* (Springer NY, New York, 1999)
68. M.V. Smolukhowski, Zs. Phys. **17**, 585 (1916)
69. B.M. Smirnov, EPL **99**, 13001 (2012)
70. B.M. Smirnov, Phys. Usp. **57**, 1041 (2014)
71. *Handbook of Chemistry and Physics*, ed. by D.R. Lide, 86 edn. (CRC Press, London, 2003–2004)
72. J.E. Hansen, D. Johnson, A. Lacis et al., Science **213**, 957 (1981)
73. J.E. Hansen, R. Ruedy, M. Sato, K. Lo, http://www.data.giss.nasa.gov/gstemp/2011
74. J. Hansen, M. Sato, R. Ruedy, http://www.columbia.edu/~jeh1/mailing/2014/20140121-Temperature2013
75. J. Hansen, M. Sato, R. Ruedy et al., http://www.columbia.edu/~jeh1/mailing/2016/20160120-Temperature2015
76. J.E. Hansen, R. Ruedy, J. Glascoe, M. Sato, J. Geophys. Res. **104**, 997 (1997)
77. J.E. Hansen, R. Ruedy, M. Sato et al., J. Geophys. Res. **106**, 947 (2001)
78. J.E. Hansen, R. Ruedy, M. Sato, K. Lo, Rev. Geophys. **48**, RG4004 (2010)
79. https://www.ncdc.noaa.gov/sotc/global/201805
80. https://www.ncdc.noaa.gov/sotc/global/201905
81. https://en.wikipedia.org/wiki/Paleontology

82. J.R. Petit, J. Jouzel, D. Raynaud, Nature **399**, 429 (1991)
83. https://www.co2.earth/21-co2-past
84. M. Milankovich, *Theorie Mathematique des Phenomenes Thermiques produits par la Radiation Solaire* (Gauthier-Villars, Paris, 1920)
85. M. Milankovich, *Canon of Insolation and the Ice Age Problem* (Belgrade, 1941)
86. http://climatica.org.uk/climate-science-information/long-term-climate-change-milankovitch-cycles
87. https://skepticalscience.com/Milankovitch.html
88. https://www.universetoday.com/39012/milankovitch-cycles
89. https://en.wikipedia.org/wiki/Solar-irradiance/media/File:Milankovitch
90. G.A. Zielinsky et al., Science **264**, 948 (1994)
91. L.I. Dorman, Ann. Geophys. **30**, 9 (2012)
92. N. Marsh, H. Svensmark, Space Sci. Rev. **94**, 215 (2000)
93. H. Svensmark, Space Sci. Rev. **93**, 175 (2000)
94. L.I. Dorman, Ann. Geophys. **23**, 3003 (2005)
95. L.I. Dorman, Adv. Space Rev. **35**, 496 (2005)
96. M.G. Ogurtsov, O.M. Raspopov, Geomagn. Aeron. **51**, 275 (2011)
97. A. Zecca, L. Chiari, J. Atmos. Sol.-Terr. Phys. **71**, 1766 (2009)
98. E.A. Kasatkina, O.I. Shumilov, M. Krapiec, Adv. Geosci. **13**, 25 (2007)
99. E.A. Kasatkina, O.I. Shumilov, N.V. Lukina et al., Dendrochronologia, **24**, 131 (2000)
100. E.P. Ney, Nature **183**, 451 (1959)
101. H. Svensmark, E. Friis-Christensen, J. Atmos. Terr. Phys. **59**, 1225 (1997)
102. H. Svensmark, Phys. Rev. Lett. **81**, 5027 (1998)
103. H. Svensmark, Proc. R. Soc. **A463**, 385 (2007)
104. H. Svensmark, T. Bondo, J. Svensmark. Geophys. Res. Lett. **36**, L151001 (2009)
105. H. Svensmark, M.B. Enghoff, J.O.P. Pedersen, Phys. Lett. A **377**, 2343 (2013)
106. J.E. Kristjansson, A. Staple, J. Kristiansen. Geophys. Res. Lett. **29**, 2107 (2002)
107. P. Laut, J. Atm. Sol.-Terr. Phys. **65**, 801 (2003)
108. N.J. Shaviv, Phys. Rev. Lett. **8**, 051102 (2002)
109. A.D. Erlykin, G. Gyalai, K. Kudela, J. Atmos. Sol.-Terr. Phys. **71**, 823 (2009)
110. A.D. Erlykin, G. Gyalai, K. Kudela, J. Atmos. Sol.-Terr. Phys. **71**, 1794 (2009)
111. M. Calisto, I. Usoskin, E. Rozanov, T. Peter, Atmos. Chem. Phys. **11**, 4547 (2011)
112. A.D. Erlykin, A.W. Wolfendale, J. Atmos. Sol.-Terr. Phys. **73**, 1681 (2011)
113. T. Sloan, A.W. Wolfendale, Environ. Rev. Lett. **8**, 0450227 (2013) **CO2**
114. https://en.wikipedia.org/wiki/Climate-sensitivity
115. https://www.co2.earth/monthly-CO2
116. https://cdiac.ess-dive.lbl.gov/ftp/trends/co2/maunaloa-co2
117. ChD Keeling, Tellus **12**, 200 (1960)
118. C.D. Keeling, R.B. Bacastow, A.E. Bainbridge, Tellus **28**, 538 (1976)
119. R.B. Bacastow, ChD Keeling, T.P. Whorf, J. Geophys. Res. **90**, 10529 (1985)
120. C.D. Keeling, T.P. Whorf, M. Wahlen, J. van der Plicht, Nature **375**, 666 (1995)
121. ChD Keeling, J.F.S. Chin, T.P. Whorf, Nature **382**, 146 (1996)
122. ChD Keeling, Ann. Rev. Energy Environ. **23**, 25 (1998)
123. https://www.esrl.noaa.gov/gmd/ccgg/trends/mlo.html
124. F. Joos, R. Spahni, Proc. Nat. Acad. Sci. USA **105**, 1425 (2008). **Energetic balance**
125. *Understanding Climate Change* (National Academy of Sciences, Washington, 1975)
126. B.M. Smirnov, *Introduction to Plasma Physics* (Mir, Moscow, 1977)
127. B.M. Smirnov, *Energetics of Atmosphere* (Moscow, Znanie, Phys. Series N3, 1979; in Russian)
128. J.T. Kiehl, K.E. Trenberth, Bull. Am. Meteorol. Soc. **78**, 197 (1997)
129. K.E. Trenberth, J.T. Fasullo, J.T. Kiehl, Bull. Am. Meteorol. Soc. **90**, 311 (2009)
130. D. Martyn, *Climates of the World* (Elsevier, Amsterdam, 1992)
131. W.F. Ruddiman, *Earth's Climate: Past and Future* (Freeman, New York, 2000)
132. S.R. Weart, *The Discovery of Global Warming* (Harvard University Press, Harvard, 2003)

133. R.T. Pierrehumbert, *Principles of Planetary Climate* (Cambridge University Press, New York, 2010)
134. D. Archer, *Global Warming: Understanding the Forecast* (Wiley, New York, 2012)
135. M. Wendisch, P. Yang, *Theory of Atmospheric Radiative Transfer* (Wiley, Singapore, 2012)
136. F. Vignova, J. Michalsky, T. Stoffel, *Solar and Infrared Radiation Measurements* (CRC Press, Boca Raton, 2012)
137. M.F. Modest, *Radiative Heat Transfer* (Elsevier, Amsterdam, 2013)
138. O. Boucher, *Atmospheric Aerosols. Properties and Climate Impacts* (Springer, Dordrecht, 2015)
139. D.L. Hartmann, *Global Physical Climatology* (Elsevier, Amsterdam, 2016)
140. G.R. North, K.-Y. Kim, *Energy Balance Climate Models* (Wiley, Weinheim, 2017)
141. H.N. Pollack, S.J. Hunter, R. Johnson, Rev. Geophys. **30**, 267 (1997)
142. J.T. Fasullo, K.E. Trenberth, Science **338**, 792 (2012)
143. K.E. Trenberth, J.T. Fasullo, Surf. Geophys. **33**, 413 (2012)
144. http://en.wikipedia.org/wiki/Earth's-energy-budget
145. https://en.wikipedia.org/wiki/Greenhouse-effect
146. C. Le Quere et al., Earth Syst. Sci. Data **10**, 2141 (2018)
147. https://www.carbonbrief.org/analysis-fossil-fuel-emissions-in-2018-increasing-at-fastest-rate-for-seven-years

Chapter 3
Water Condensation Processes in Atmospheric Air

Abstract Cloud properties and cloud processes are analyzed in this chapter. Clouds consisting of water microdroplets are formed in atmospheric condensation processes and partake in the water circulation through the atmosphere where water evaporates from the land or oceans and returns to the Earth's surface in the form of precipitations mostly. Because of a restricted humidity of the standard atmosphere at any altitudes, water condensation proceeds there under the action of vertical winds which mix low wet layers of the atmosphere with cold high ones. The mechanisms of the formation and growth of water microdroplets in the atmosphere include processes of evaporation of water molecules from microdroplets, attachment of free water molecules to them, coagulation and coalescence processes, and gravitation growth of droplets which fall down under the action of their weight. Usually, kinetics of the growth of water microdrops in clouds results from the competition between the coalescence and gravitation growth mechanisms. A high density of free and bound water molecules is realized in cumulus clouds where, due to the equilibrium between free water molecules and microdroplets which is established fast, the number density of free water molecules is equal to that at the saturated vapor pressure. In addition, the transformation of neutral microdroplets in water drops of precipitation proceeds fast compared with the lifetime of cumulus clouds. Hence, water microdroplets are charged in cumulus clouds and their charges have the same sign. Thus, processes of growth of water droplets in cumulus clouds proceed simultaneously with electric processes. As a result, processes of growth of water microdrops in cumulus clouds are connected with atmospheric electricity.

3.1 Water Microdrops in Water Circulation

3.1.1 Clouds in Troposphere

Condensed water is located in the atmosphere in the form of clouds which consist of microdroplets or water particles of micron sizes. This means that water particles are not distributed more or less uniformly in the atmosphere, but are concentrated in a

B. M. Smirnov, *Global Atmospheric Phenomena Involving Water*,
Springer Atmospheric Sciences, https://doi.org/10.1007/978-3-030-58039-1_3

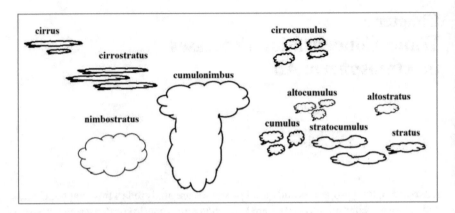

Fig. 3.1 Types of clouds depending on the altitude [1]

small part of the atmosphere. This results from the character of water condensation in the atmosphere, and we consider these processes below. Because the main part of condensed water is located in the liquid state, we consider below only liquid water particles or water microdroplets. We first represent types of clouds which consist of microdroplets and are given in Fig. 3.1.

Classification of cloud types was represented firstly by Howard [3, 4] at the beginning of the nineteenth century. Though his classification is problematic, it is a basis of contemporary definition of cloud types. In this classification, there are three cloud shapes, namely, cirrus, cumulus, and stratus ones. Cirrus means a curl or hair. Such clouds contain rare condensed water and are present at high altitudes. Stratus clouds are characterized by a higher water density and usually they cover all of the sky or most of its parts. Stratus clouds may consist of several tiers. The water density is higher in cumulus clouds which have the form of separated heaps. These types of clouds are represented in Figs. 3.2, 3.3 and 3.4.

Clouds may have mixed forms. Some of them are given in Fig. 3.1. In addition, the prefix "alto" relates to clouds located at high altitudes, and the suffix "nimbus" corresponds to dense clouds. Such clouds may be a source of rain, i.e. rain may be created from cumulonimbus clouds and nimbostratus clouds which are represented in Fig. 3.1. In addition, dense cumulonimbus clouds can cause thunderstorm, and Fig. 3.5 shows an example of a cumulonimbus towered cloud. In this case, clouds are formed under the action of rising streams of wet air, and condensation takes place in a wide range of altitudes. In addition, water microdrops are charged that leads to electric phenomena including these clouds. The accumulation of condensed water at high altitude creates a towered shape of the cloud under the action of rising air streams.

Clouds are the object of this investigation as an element of atmospheric physics and may be examined from various standpoints [8–24]. Being guided by electric and radiative processes in the atmosphere, in which clouds participate, we consider clouds as an ensemble of water microdroplets. These microdroplets along with bound water

Fig. 3.2 Cirrus clouds [2]

Fig. 3.3 Stratus clouds [5]

Fig. 3.4   Cumulus clouds [6]

Fig. 3.5   A cumulonimbus cloud [7]

molecules may include various additive materials which determine the cloud color, i.e. their absorption in a visible spectrum range, where liquid water is transparent. Besides that, atmospheric microparticles—aerosols may consist of other materials, and they are the object of special consideration [25–28]. Below for simplicity, we are basing on a simple cloud model, where a cloud consists of water microdroplets of an identical size, and these droplets are found in the liquid aggregate state. In other words, within the framework of the model under consideration, clouds consist of liquid microdroplets of the same radius. Moreover, because parameters of these microdroplets are different for different types of clouds, we are guided by dense clouds which are of importance for the phenomena under consideration. In particular, by analyzing electric properties of the atmosphere, we are restricted by cumulus clouds, where average parameters of water microdroplets are as follows [29–32]:

$$r = 8\,\mu m, \quad N_d = 10^3\,cm^{-3}. \tag{3.1}$$

We assume here an individual microdroplet to have a spherical shape, r is a droplet radius, and N_d is the number density of droplets.

A number of bound water molecules of an individual microdroplet of a cumulus cloud and its radius r are connected by the relation

$$n = \left(\frac{r}{r_W}\right)^3, \tag{3.2}$$

where r_W is the Wigner–Seitz radius [33, 34] which is equal to $r_W = 1.92\,\text{Å}$ [35] for liquid water. According to this formula, the number of water molecules in a water droplet of radius $r = 8\,\mu m$ is equal to $n = 7 \times 10^{13}$, and the droplet mass is $m_o = 2.1 \times 10^{-9}$ g. The average number density of bound water molecules in microdroplets is $nN_d = 7 \times 10^{16}\,cm^{-3}$. As is seen, the concentration of bound water molecules in cumulus clouds has the same order of magnitude as the average concentration of free water molecules in the atmosphere.

The parameter which characterizes the degree of covering of the sky by clouds is cloudiness [37] which is the sky part that is covered by clouds. It is clear that this parameter is definite, if the optical thickness of clouds is large in the visible spectrum range for the opaque atmosphere. If we take the limited optical thickness to be 0.1, the average clouding is equal to 0.68 in this case [37]. Then, one can extract three basic tiers where clouds are located. The evolution of cloudiness for these three tiers are represented in Fig. 3.6 along with the total water amount in clouds. In addition, the total cloudiness over the land is 0.10–0.15 less than that over oceans [37].

If we take the limiting optical thickness to be 2 for clouds which account for the cloudiness, the latter is equal to 0.56. In this case, cirrus clouds are accepted as transparent and are excluded from the consideration. Of course, cloudiness depends on the latitude and seasons. But in the above analysis, we deal with global parameters, i.e. with averaged parameters over the globe.

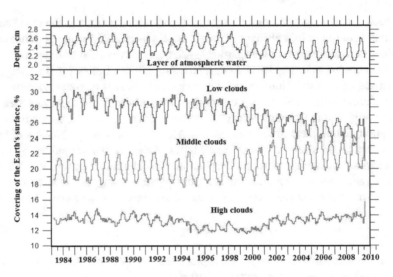

Fig. 3.6 Evolution in time for cloudiness according to data of NOAA Earth System Laboratory (NASA) [36]. Above, the total amount of atmospheric water which results from the transformation of atmospheric water in the liquid one and a uniform distribution of liquid water over the Earth's surface is given

3.1.2 Water Circulation Through the Atmosphere

As a result of the equilibrium between atmospheric and surface, water evaporates from the Earth's surface and returns back in the form of precipitation. A general character of this equilibrium between atmospheric and surface water is called the water circulation in nature [38], and its scheme is represented in Fig. 3.7. Because the rate of evaporation is large compared to that for the transport of water droplets to the Earth's surface, the mass of condensed water of the atmosphere is small compared with that in the form of free molecules.

One can expect the following scheme of water circulation given in Fig. 3.7. Namely, water molecules which evaporate from the Earth's surface rise up due to the diffusion in atmospheric air, and there in cold layers water is condensed. Then, water molecules join in the microdroplets which, in turn, form rain droplets. But within the framework of the model of standard atmosphere, another scenario of processes takes place. In the first approach, only 1 and 2 processes of Fig. 3.7 partake in this phenomenon, i.e. evaporated water molecules are removed from the Earth as a result of convective diffusion (process 1 of Fig. 3.7) and they return to the Earth under the action of the gravitation force (process 2 of Fig. 3.7) which establishes an equilibrium in this system. In this scheme, other processes are weak.

The circulation mechanism under consideration is accompanied by the specific character of dynamics of atmospheric air which contains clouds [17, 21, 39–41]. The chain of circulation processes of atmospheric water involves processes of formation

Fig. 3.7 Sequence of processes in the course of water circulation in the Earth's atmosphere. 1—evaporation and convective diffusion of water molecules from the Earth's surface, 2—drift of water molecules to the Earth's surface under the action of the gravitation force, 3—condensation of evaporated water and horizontal displacement of forming droplets under the action of winds, 4—gravitation falling of water microdrops, 5—rain, 6—lightning

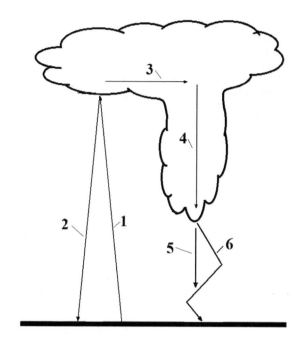

and transport of condensed water. In the course of air circulation, water molecules are captured by air vortices which provide air transport, whereas these vortices do not act on water microdroplets.

In reality, certain conditions are required to provide water circulation in the atmosphere. Low-scale circulation is created as a result of a nonuniform landscape. For example, mountains cause the arising of vertical air fluxes, and a neighborhood of a lake and land with a different character of heating of the Earth's surface under the action of solar radiation creates a vertical flux also that equilibrates these fluxes.

Along with a low-scale atmospheric circulation which is considered above, a large-scale circulation is created by a different character of the heat balance for different globe parts. Air circulation in the atmosphere occurs under the action of forces resulting from nonuniform heating of the surface. The corresponding horizontal gradients occur at both large and small scales. Figure 3.8a [42] explains the occurrence of large-scale temperature gradients which cause a long-distance air motion. Processes of the energy balance of the atmosphere and the Earth due to absorption and reflection of solar radiation, as well as atmospheric emission and absorption of infrared radiation, violate the local thermodynamic equilibrium and create temperature gradients. These gradients cause by the movement of atmospheric air, as shown in Fig. 3.8b [43], so that warm air from the region with an excess heating is replaced by cold air from regions of insufficient heating.

In particular, Fig. 3.8b represents large-scale circulations of atmospheric air [44–46] which include the Hadley cell as the main large-scale air currents along the corresponding meridians. Along with this, there is a large number of less large-scale

Fig. 3.8 Large-scale circulation in the atmosphere. **a** Character of the Earth's energetic balance which causes large-scale air transport [42]. **b** Large-scale air flows [43]. Blue arrows correspond to motion of cold air flows, red arrows relate to hot air flows

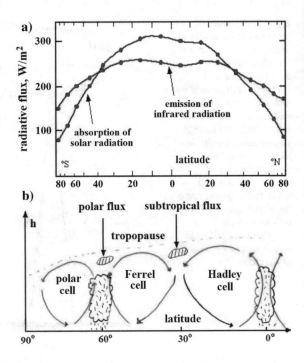

stable air movements that are associated with the landscape, located on the surface of oceans and mountains. The terrain can also cause temporary air currents for the same reason, such that the overheating of the Earth's surface in a certain area causes air movements in the form of wind, which maintains the energy balance. As can be seen, in these cases, the movement of air occurs in the form of vortices of different scales. As a result of these air currents, the abovementioned mixing of warm wet air from the Earth's surface with cold air of the high layers of the atmosphere also occurs, which is accompanied by the condensation of water vapor in atmospheric air with formation of clouds.

3.2 Formation of Clouds Under Equilibrium Conditions

3.2.1 Conditions of Water Condensation in Atmospheric Air

In considering the processes of water condensation in the atmosphere, we take into account a small time of evaporation of droplets in the case of their small size. This means a small time of equilibrium establishment between free water molecules into the atmosphere and droplets of a small size, and we check it further on the basis of rates of water condensation and evaporation under indicated conditions; water

Table 3.1 Parameters of a saturated water vapor in atmospheric air [48, 49]: p_{sat} is the water saturated vapor pressure at an indicated temperature, N_{sat} is the number density of water molecules in a saturated vapor, A_s is the ratio of a mass of a saturated water vapor to an air mass located in a given volume at atmospheric pressure

T, K	p_{sat}, Torr	N_{sat}, 10^{17} cm^{-3}	A_s, g/kg
253	0.77	0.295	0.631
258	1.24	0.463	1.01
263	1.95	0.716	1.59
268	3.02	1.09	2.46
273	4.58	1.62	3.74
278	6.55	2.27	5.35
283	9.22	3.14	7.52
288	12.8	4.29	10.4
293	17.6	5.78	14.3
298	23.8	7.70	19.4

droplets may exist stably in atmospheric air if the air temperature is below the dew point [47], that is the temperature at which the water partial pressure in air is equal to the saturated vapor pressure. Table 3.1 contains the saturated vapor pressure and corresponding number density of molecules depending on the air temperature.

It is convenient to approximate the number density of water molecules at the saturated vapor pressure $N_{sat}(T)$ as a temperature T function by the dependence

$$N_{sat}(T) = N_o \cdot \exp(-E_{sat}/T), \tag{3.3}$$

Parameters of this dependence according to data of Table 3.1 are $N_o = 1.1 \times 10^{26}$ cm^{-3} and $E_{sat} = 0.48$ eV.

Since the water condensed phase exists in air stably, if the partial pressure of water molecules exceeds or is equal to the saturated vapor pressure, the condition of the presence of the condensed phase in air has the form

$$N_w \geq N_{sat}, \tag{3.4}$$

where N_w is the total number density of free water molecules. For the model of standard atmosphere, the number density of water molecules near the Earth's surface is $N_w = 4.3 \times 10^{17}$ cm^{-3} that corresponds to the saturated number density at the global temperature according to the data of Table 3.1.

In this analysis, it is convenient to operate also with the atmospheric moisture η which is defined by formula (2.4) and is the ratio of the number density of free water molecules in the atmosphere to that of the saturated vapor pressure for a given temperature. The criterion (3.3) of water condensation corresponds to a supersaturated vapor, where

$$\eta \geq 100\% \tag{3.5}$$

But this criterion is not fulfilled for the model of standard atmosphere. According to data of Fig. 2.2, it is approximately 82%, i.e. we deal with a non-saturated atmospheric water vapor near the Earth's surface. Thus, we come to the contradiction, so that within the framework of the model of standard atmosphere the number density of water molecules near the Earth's surface is $N_w = 4.3 \times 10^{17}$ cm^{-3} that corresponds to the saturated vapor $\eta = 100\%$, whereas the average moisture at the Earth's surface is 82% according to the data of

Fig. 2.2. This discrepancy determines the accuracy of us ed approximations and averaging. In particular, one can see the accuracy of approximation of certain measurements of Figs. 2.3, 2.4, and 2.5 by the dependence (2.5) is characterized by the accuracy of (20–30)%.

Let us analyze the altitude dependence for the saturated number density N_{sat} of water molecules. From formula (3.3), we have

$$N_{sat}(h) \sim \exp\left(-\frac{h}{h_{sat}}\right), \quad h_{sat} = \frac{T^2}{E_{sat} \cdot |dT/dh|} \approx 2.3\,\text{km}, \tag{3.6}$$

where we take the air temperature near the Earth's surface $T = T_E = 288$ K and the temperature gradient $dT/dh = -6.5$K/km for the model of standard atmosphere. Comparing the altitude dependencies (2.5) and (3.6), one can conclude that the moisture decreases on average with an increasing altitude, at least, in the lower layers of the troposphere. We use here the temperature gradient for the model of standard atmosphere.

One can see that the number density of water molecules in an atmospheric layer is restricted by their value (3.3) at the saturated vapor pressure for the temperature of this layer. From this, one can determine the maximum mass of a water vapor in the atmosphere in the form of free water molecules. Taking the maximum number density of water molecules $N_w = N_{sat}$ for the model of standard atmosphere, one can obtain the maximum water mass per unit area of the Earth's surface as

$$\rho_{max} = N_s h_{sat} m_o \approx 3.0\,\text{g/cm}^2, \tag{3.7}$$

where the number density of water molecules near the Earth's surface is $N_s = 4.3 \times 10^{17}$ cm^{-3} in accordance with Table 3.1 for standard atmosphere. If we transfer this water in the liquid state and distribute uniformly over the Earth's surface, the average height of a formed water layer will be 3.0 cm. Note that the transformation of total atmospheric water of standard atmosphere into liquid and uniform distribution of atmospheric water over the globe lead to the height of $h = 2.5$ cm for a formed layer. One can see that the mass of atmospheric water in the form of free molecules is close to its limiting value. Indeed, because formula (3.7) gives the density of the saturated water vapor in atmospheric air, from this it follows that the average atmospheric moisture is approximately 80%.

Thus, in the atmosphere under equilibrium conditions which are described by the model of standard atmosphere and the equilibrium is supported by convective motion of the atmosphere, the moisture is below 100% and the condensation process is absent. Hence, the water condensed phase may be formed in atmospheric air only as a result of fluctuations in the atmosphere or by nonequilibrium processes.

3.2.2 Formation of the Condensed Phase resulted from Mixing of Atmospheric Layers

We obtain above that the formation of condensed water phase in the atmosphere requires nonequilibrium conditions. We consider below such a real situation where some part of the Earth's surface under the influence of solar radiation is heated more strongly than the neighboring areas This causes a thermal instability in the form of ascending air flows in this area of the surface and descending flows in neighboring areas. This movement of air leads to the mixing of air layers, so that warm wet air layers penetrate into high cold layers where atmospheric water becomes supersaturated at the temperature of these layers.

Focusing on this character of air transport in the atmosphere, we consider below adiabatic mixing of two volumes of atmospheric air taken from different layers with parameters corresponding to the model of the standard atmosphere. Our task is to determine the conditions at which the dew point can be reached, i.e. the air humidity reaches 100%. Along with this, we determine the maximum possible supersaturation of an atmospheric water vapor under given conditions.

So, we fulfill the standard operation of mixing two volumes (parcels) of atmospheric air, where the first volume is located at the beginning near the Earth's surface, and the second one is taken from an altitude of h. Mixing of the layers occurs adiabatically, so that heat does not go beyond the mixed volumes. Under these conditions, the temperature of the mixture is equal to

$$T = \frac{n_1 T_1 + n_2 T_2}{n_1 + n_2} = T_1 - (T_1 - T_2)x, \ x = \frac{n_2}{n_1 + n_2}, \tag{3.8}$$

where n_1, n_2 are the numbers of air molecules in each of these parcels, T_1, T_2 are temperatures of these layers at the beginning, and x is the concentration of air molecules taken from the second parcel. At the same time, we assume the air heat capacity to be independent of the temperature, and take into account the adiabatic character of parcel mixing. Then the mixture temperature is given by

$$T = \frac{n_1 T_1 + n_2 T_2}{n_1 + n_2} = T_1 + hx\frac{dT}{dh}, \tag{3.9}$$

where $dT/dh = -6.5\,\text{K/km}$ is the temperature gradient for standard atmosphere.

In considering the character of mixing of two parcels, we assume the number density of water molecules in the near-surface layer to be varied depending on the moisture, but the number density of water molecules in the second parcel we take due to the standard atmosphere model in accordance with formula (2.5). Correspondingly, the number density of water molecules of the mixture

$$N_w = \eta N_{\text{sat}}(T_E)(1 - x) + N_o \exp(-h/\lambda)x \tag{3.10}$$

Here, η is the air moisture near the Earth's surface; we take the temperature of a lower parcel as the global one $T_2 = T_E = 288$ K, and the number density of water molecules is given by formula (2.5). If this number density exceeds $N_{\text{sat}}(T)$ that corresponds to the saturated vapor pressure, the water vapor becomes a supersaturated one, and a part of water molecules transfers into the condensed phase in a space.

Let us construct the function $\Phi(h, x)$ as

$$\Phi(h, x) = \frac{N_w}{N_{\text{sat}}(T)} = \frac{\eta N_{\text{sat}}(T_E)(1 - x) + N_o \exp(-h/\lambda)x}{N_{\text{sat}}(T_E - h(1 - x)\frac{dT}{dh})} \tag{3.11}$$

This function characterizes the degree of vapor saturation. The saturation threshold is described by the equation $\Phi(h, x) = 1$. We take the below within the accuracy of parameters in expression (3.11) according to formulas (2.5) and (3.3) $N_o = N_{\text{sat}}(T_E)$ for the model of standard atmosphere. This allows one to reduce expression (3.11) to the form

$$\Phi(h, x) = \left[\eta(1 - x) + x \exp\left(-\frac{h}{\lambda} \right) \right] \exp\left[\frac{E_{\text{sat}}}{T_E - h(1 - x)dT/dh} - \frac{E_{\text{sat}}}{T_E} \right], \tag{3.12}$$

where $T_E = 288$ K, $E_{\text{sat}} = 0.48$ eV, $\lambda = 2$ km.

One can see that if the condition $\Phi(h, x) > 1$ holds true, air is supersaturated and admits the formation of the condensed water phase. It is convenient to introduce the supersaturation degree

$$S(h, x) = \Phi(h, x) - 1, \tag{3.13}$$

and equation $S(x, h) = 0$ characterizes the threshold of water condensation in atmospheric air, and formation of the water condensed phase is possible under the criterion $S(x, h) > 0$ for the equilibrium between water aggregate states. It is clear that the supersaturation as a result of air transfer from the Earth's surface to higher layers starts from some altitudes. These altitudes are given in Fig. 3.9a for some parameters of mixing. Figure 3.9b contains the degree of supersaturation for the air mixture.

It should be noted that the above analysis is a demonstration of the condensation possibility in atmospheric air, rather than the description of the condensation process. From this it follows that condensation is impossible in motionless air of standard atmosphere [50] since an atmospheric water vapor is nonsaturated there. But because of a high average atmospheric moisture, air condensation is realized as a result of

Fig. 3.9 Character of adiabatic mixing of two parcels, where the first one contains wet air taken from the Earth's surface and the second one is located at an altitude h with parameters of standard atmosphere. **a** Minimal atmospheric altitude at which the dew point of air from mixed layers may be reached. **b** The degree of supersaturation $S(h, x)$ which is attained at an indicated altitude

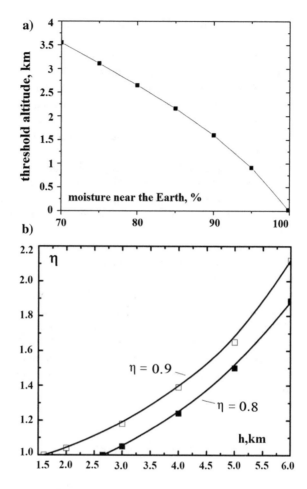

the mixing of layers with different temperatures. Therefore, convective transport of air in the atmosphere [51–53] is of importance for the condensation process which proceeds in cold atmospheric regions. In spite of energy release in the course of water condensation, this process intensifies as a result of transport phenomena, and condensing water is concentrated in restricted atmospheric regions [41, 54, 55]. Since this process is determined by convective transport in the atmosphere, the creation of cumulus clouds is a random process which is accompanied by large gradients of the water density. According to the existing models of water transport in clouds (for example, [51, 56–59]), convective transport leads both to the creation and smearing of cumulus clouds.

Thus, from the above analysis it follows that clouds are formed starting from some altitudes above the Earth in accordance with observations. Of course, a fog can be formed near the Earth's surface if the temperature of near-surface layers is low. Such conditions may be realized, for example, at autumn tomorrow, but formed condensed water in this case does not partake in circulation of atmospheric water, as well as in electric processes of the atmosphere. Therefore, we restrict above by condensation processes in a cumulus cloud which proceed as a result of mixing of air jets under the action of vertical winds. In this case, condensation processes proceed starting from some altitudes and include restricted portions of the atmosphere. Hence, the amount of condensed water in the atmosphere is less compared to that consisting of free molecules. Indeed, let us take a cumulus cloud with average parameters (3.1) where the number density of bound water molecules in microdroplets is $nN_d = 7 \times 10^{16}$ cm^{-3}. If this cloud is located at the altitude of 3 km, where the temperature is 268 K for standard atmosphere, we have for the number density of free water molecules $N_w = 1.1 \times 10^{17}$ cm^{-3} according to Table 3.1. As is seen, in this case the number density of free water molecules exceeds that of the bound ones. In addition, regions with such condensed water include a small atmosphere part.

One can expect that the system consisting of air with water microdroplets and free water molecules at the saturated vapor pressure is unstable in the atmosphere. Indeed, let us take an element of a cumulus cloud with average parameters (3.1) and displace it in a lower layer of standard atmosphere. According to Table 3.1, at the temperature 274 K condensed water disappears if the equilibrium is supported in atmospheric air between free and bound water molecules. Evaporation of condensed water in this case requires an energy which is able to heat air with this water by 6 K. Because the process of microdroplet evaporation requires an additional energy, this process increases the stability of the mixture of free and bound water in atmospheric air.

3.2.3   Rate of Water Condensation

Let us formulate a general character of atmospheric processes which result in water circulation. These processes include atmospheric motion under the action of gravitation forces and gradients of the density of atmospheric compounds, because these factors create simultaneously the convection motion of atmospheric air, water condensation, as well as the altitude distribution for air and water molecules. Now our task is to estimate the water flux in the atmosphere from the Earth's surface and opposite fluxes where water may be found in the form of free molecules as well as the water condensed phase. Being guided by the model of standard atmosphere, we deal with stationary fluxes of air and water both from the Earth's surface and in the inverse direction. Because of an irregular character of motion of atmospheric air, one must consider resulting values of fluxes within the stationary model as estimations. In these evaluations, we take into account that the altitude profiles for air and water molecules give information about the flux of bound molecules to the Earth.

We first determine the flux of water molecules from the Earth's surface which evaporate from it. Taking the total rate of evaporation $J_w = 1.5 \times 10^{13}$ g/s, we have for the flux of water molecules from the surface

$$j_w = \frac{J_w}{m_o S} = 9 \times 10^{16} \, \text{cm}^{-2} \, \text{s}^{-1}, \tag{3.14}$$

where $S = 5.1 \times 10^{18}$ cm^2 is the area of the Earth's surface, and $m_o = 3.0 \times 10^{-23}$ g is the mass of an individual water molecule. One can take the same value on the basis of the height of liquid water $h_w = 2.5$ cm which results from the transformation of atmospheric water in the liquid aggregate state and this water is distributed uniformly over the Earth's surface. Introducing the residence time τ of water molecules in the atmosphere, one can obtain for the fluxes of water molecules for the formation and decomposition of atmospheric water

$$j_w = \frac{\rho h_w}{m_o \tau_w} = 1.1 \times 10^{17} \, \text{cm}^{-2} \, \text{s}^{-1}, \tag{3.15}$$

where $\rho = 1$ g/cm^3 is the density of liquid water, and $\tau_w = 9$ days is the residence time of a water molecule in the atmosphere. The degree of coincidence of fluxes according to formulas (3.14) and (3.15) testifies the accuracy of values for used parameters.

We now determine the flux of atmospheric water molecules which follows from the space distribution (2.5) of water molecules. We have

$$j_w = -D_L \frac{dN_w}{dh} = \frac{D_L N_w}{\lambda} = 9 \times 10^{16} \, \text{cm}^{-2} \, \text{s}^{-1}, \tag{3.16}$$

where we use the average number density of water molecules near the Earth's surface as $N_w = 4.3 \times 10^{17}$ cm^{-3}, the effective diffusion coefficient due to convection motion is taken as $D_L = 4 \times 10^4$ cm^2/s according to formula (2.10), and also $\lambda = 2$ km. In this consideration, we assume the convective character of air motion, i.e. this motion is a sum of vortices of some sizes below 1 m. Therefore, the transport for large distance has a diffusion character, and according to an estimation we have $D_L \sim Lv$, where $L \sim 1$ m is a maximal size of vortices, $v = (1\text{--}10)$ m/s is a typical velocity of atmospheric jets and winds. One can see that the diffusion coefficient D_L depends on certain atmospheric conditions, and the used value is characterized by an accuracy of approximately 20%, as well as results of these evaluations.

Note that formulas (3.14), (3.15), and (3.16) characterize fluxes of water molecules from the Earth's surface on the basis of various considerations. In considering water circulation through the atmosphere, one can determine water fluxes to the Earth's surface due to free water molecules j_m and to condensed atmospheric water j_c. Evidently, from the conservation of the total flux, we have

$$j_w = j_m + j_c \tag{3.17}$$

If we neglect the condensation process, water molecules are captured by air vortices, and their falling to the Earth, due to the convection process and under the action of the gravitation field, is equal to

$$w_m = \frac{D_L}{\Lambda} = 0.05\,\text{cm/s} \tag{3.18}$$

This value relates both to air molecules and to water ones which move to the Earth under the action of the gravitation field. Formula (3.18) accounts for the interaction of convection motion of air with captured water molecules with their falling to the Earth. Evidently, this character of motion toward the Earth's surface will be conserved for water molecules if condensation takes place at altitudes far from the surface. Then the flux of free water molecules to the Earth's surface j_m is equal to

$$j_m = N_w w_m = \frac{D_L N_w}{\Lambda} = 2 \times 10^{16}\,\text{cm}^{-2}\,\text{s}^{-1} \tag{3.19}$$

From this on the basis of the above formulas and the balance of water fluxes, we have for the flux of condensed atmospheric water near the Earth's surface

$$j_c = j_w - j_m = 7 \times 10^{16}\,\text{cm}^{-2}\,\text{s}^{-1} \tag{3.20}$$

As is seen, the water flux results from the Earth due to precipitations of water molecules or due to the condensed phase in the form of falling water droplets. This confirms the initial version of the water circulation through the atmosphere according to which water penetrates in the atmosphere in the evaporation processes in the form of free water molecules and water leaves the atmosphere as a result of precipitation.

From this it follows also that due to the condensed phase of atmospheric water, the drift velocity of water to the Earth's surface is larger than that given by formula (3.18). Indeed, the effective drift velocity w_{ef} of atmospheric water toward the Earth is equal to

$$w_{ef} = \frac{D_L}{\lambda} = 0.2\,\text{cm/s} \tag{3.21}$$

One can compare the average flux (3.20) of atmospheric condensed water in the vicinity of the Earth's surface with that j_d due to falling of water microdroplets in cumulus clouds. Taking typical water parameters (3.1) in a cumulus cloud, according to which the average number density of bound water molecules is $N_b = n N_d = 7 \times 10^{16}\,\text{cm}^{-3}$, and the falling velocity which according to formula (2.22) is equal to $w_g = 1\,\text{cm/s}$, we have

$$j_d = N_b w_g = 7 \times 10^{16}\,\text{cm}^{-2}\,\text{s}^{-1} \tag{3.22}$$

Note that j_c and j_d correspond to different quantities, such that j_c describes the average flux of free water molecules located near the Earth's surface, while j_d is the flux of bound molecules due to water microdroplets at some altitudes above the Earth. The latter one occurs only in cumulus clouds which cover only a part of the globe's surface. In the course of motion of this flux to the Earth's surface, it is increasing due to the attachment of free water molecules to water droplets. From this it follows only that $j_c \sim j_d$.

3.3 Kinetics of Growth of Water Microdroplets in Atmospheric Air

3.3.1 Mechanisms of Drop Growth in Air

We convince above that the formation of the condensed phase of atmospheric water is possible in a supersaturated water vapor according to (3.4). If this criterion holds true, a part of a water vapor is transformed into small particles or microdroplets. The classical character of formation and growth of a new phase in a weakly supersaturated gas under the thermodynamic equilibrium [60–63] proceeds through the origin of an embryo which is subsequently the condensation nucleus. This embryo is a water particle of a critical radius [62, 63]. The further growth of the condensed phase results from attachment of these molecules to condensation nuclei with the probability of almost one [60–64], and a typical time of particle growth up to the critical radius is large.

This process proceeds in a pure vapor if a forming particle consists of molecules of the same type, and this stage is long because the probability of decay for a cluster with a size below the critical one is close to one. Therefore, the presence in the gas so-called nuclei of condensation accelerates the growth process. Atmospheric ions or radicals may be nuclei of condensation. Therefore, in the atmosphere where nuclei of condensation exist, the growth of water particles occurs with the participation of condensation nuclei and proceeds until the vapor pressure exceeds the saturated one. Correspondingly, the process of growth of a new phase is called as the nucleation process.

There are four mechanisms for formation and growth of the condensed phase in the form of nano-clusters. These mechanisms are important for water microdroplets in atmospheric air and are presented in Fig. 3.10. In this case, we ignore a delay of the growth process due to the classical theory of nucleation in a pure gas [60–63], i.e. we assume the presence of condensation nuclei in the atmosphere which accelerate the growth process. In this consideration, we take into account the general specifics in the formation of nanometer-sized particles—aerosols, [68–71] including that the presence of a new phase influences the rate of the nucleation process [72].

Under real conditions of location of an atmospheric water vapor in atmospheric air, the growth mechanisms of Fig. 3.10 are sufficient to describe the growth of a new water phase. We restrict below by the liquid water phase only as a more spread case of processes under consideration. There are two regimes of nucleation depending on

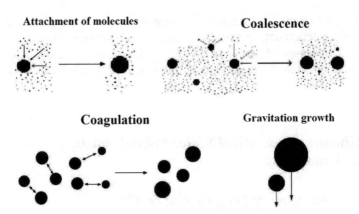

Fig. 3.10 Mechanisms of formation and growth of water droplets in atmospheric air [65–67]

the character of motion of the nucleating molecules in a surrounding gas [73–75]. We will call these regimes as kinetic and diffusion ones [65], so that in the kinetic regime of growth of a new phase, an atomic particle which partakes in the nucleation process moves along a rectilinear trajectory, whereas in the opposite diffusion regime of growth the diffusion character of motion of nucleating particles is realized.

Because of a large density of molecules in air and hence a relatively small mean free path of atomic particles in air, we restrict below ourselves by the diffusion regime of growth for nucleation processes under consideration. This means that a size of a growing water droplet is large compared to the mean free path of molecules during a basic growth stage, and in the course of this process, a nucleating droplet interacts with many air molecules simultaneously. In this case, the character of interaction of a moving droplet with air molecules is described by the air viscosity. In particular, in the case of the coagulation process represented in Fig. 3.10, the rate constant of this process is given by [67]

$$k_{as} = \frac{8T}{3\eta},\tag{3.23}$$

where η is the viscosity coefficient of atmospheric air. In particular, under atmospheric pressure and room temperature in air formula (3.23) gives $k_{as} = 5.8 \times 10^{-10}$ cm^3/s. This leads to a typical time of the order of 10^6 s for the doubling of the number of water molecules in a microdroplet of a cumulus cloud with parameters (3.1). Therefore, this mechanism of growth of water droplets in the atmosphere is not important under typical atmospheric conditions, and we ignore it below.

Let us consider the first stage of the formation and growth of the water condensed phase in the atmosphere when, in accordance with Fig. 3.10a, excess water molecules attach to a condensation nucleus. Then the evolution of the number of molecules for a test microdroplet n is given in accordance with the Smoluchowski formula [76]

$$\frac{dn}{dt} = 4\pi D_w r N_w - 4\pi D_w r N_{sat},\qquad(3.24)$$

where n is a current number of molecules in a droplet, D_w is the diffusion coefficient of molecules in the surrounding air, N_w is a current number density of free water molecules, and N_{sat} is the number density of water molecules at the saturated vapor pressure. The first term of formula (3.24) is the rate of attachment of water molecules to a test microdroplet, the second one is the rate of its evaporation, and these terms are identical at the saturated vapor pressure.

The diffusion coefficient of water molecules in the air at atmospheric pressure is $D_w = 0.22\,\mathrm{cm}^2/\mathrm{s}$ [77, 78]. If a microdroplet is inserted into atmospheric air that does not contain water vapor, its evaporation time τ_{ev} of this droplet is equal to

$$\tau_{ev} = \frac{3r^2}{4\pi D_w N_{sat} r_W^3},\qquad(3.25)$$

where $r_W = 1.92\,\text{Å}$ [35] is the Wigner–Seitz radius for water, and r is the droplet radius.

In particular, let us use parameters of a cumulus cloud which is located at an altitude of $h = 3\,\mathrm{km}$, where the temperature is $T = 268\,\mathrm{K}$ which corresponds to the number density of water molecules at the saturated vapor pressure $N_{sat} = 1.1 \times 10^{17}\,\mathrm{cm}^{-3}$. Further, the air pressure at this altitude is about 40% lower than that at the Earth's surface, which corresponds to the diffusion coefficient of water molecules in air $D_w = 0.31\,\mathrm{cm}^2/\mathrm{s}$. On the basis of these parameters, we obtain from formula (3.25) for a typical microdroplet radius $r_o = 8\,\mu\mathrm{m}$ in a cumulus cloud, and we obtain for the evaporation time of this microdroplet in a vacuum $\tau_{ev} \approx 0.6\,\mathrm{s}$. This value is small compared to a typical lifetime of clouds, as well as a time of its evolution which is measured by hours. Therefore, in consideration of the growth and evolution of water microdroplets in atmospheric air, we assume that in the course of this evolution, an equilibrium is maintained between water microdroplets and free water atmospheric molecules.

The concept of water nucleation in atmospheric air is based on a large number density N_{nc} of condensation nuclei, which provides the total transformation of excess water molecules into microdroplets. Denoting the number density of these excess molecules by N_w, we have for the average number n of molecules in one microdroplet when all excess molecules become bound ones

$$n = \frac{N_w}{N_{nc}}$$

From this, we have that for formation of a cumulus cloud with parameters (3.1) of microdroplets, the following relation for the number density of nuclei condensation must be valid

$$N_{nc} \gg 10^4\,\mathrm{cm}^{-3}\qquad(3.26)$$

Because the number density of atmospheric ions at an altitude $h = 3\,\mathrm{km}$ is lower than this value, some radicals must be nuclei of condensation.

This rough estimation allows us to represent the character of condensation of the atmospheric water vapor. Nuclei of condensation are necessary for growth of water droplets. In reality, the attachment of water molecules to nuclei of condensation is a chemical reaction between water molecules and radicals in the form of nitrogen oxides, ammonia, and sulfur compounds [79–83]. Attachment of water molecules to microdroplets proceeds with the participation of some acids and their salts. When the size of drops becomes large, the character of the first stage of growth does not influence the subsequent droplet growth.

Thus, the evolution of water microdroplets in clouds results from processes of coalescence or Ostwald ripening [84, 85] that results from the interaction between microdroplets and free water molecules. The gravitation growth of water droplets in the atmosphere is of importance for the origin of rain and relates to a large droplet size. Then, large microdroplets fall faster than small ones. Overtaking small droplets, large droplets join with them. As a result, water microdroplets are converted into millimeter drops of rain. The size distribution function of droplets which is established in the course of the coalescence process has the automodel form [86–89]. The reduced size $u = n/n_{cr}$ is the ratio of a number of drop molecules n to the critical number n_{cr}, and the automodel form of the distribution function $f(u)$ for the coalescence process [86–90] implies that the distribution function $f(u)$ does not vary in the course of coalescence.

In analyzing general principles of the coalescence process, we note its self-consistent character. On the one hand, the total flux of molecules attached to large drops and evaporated from small drops are identical. On the other hand, as a result of the total process, the average size of drops grows and their number density decreases. Simultaneously, the number density of free molecules decreases slowly. Because the number of drop molecules is large $n \gg 1$, one can use a small parameter $n^{-1/3}$ and represent the total binding energy E_n of bound molecules in the drop as an expansion over this small parameter in the form [91]

$$E_n = \varepsilon_o n - A n^{2/3}, \tag{3.27}$$

where ε_o is the average binding energy per one water molecule in a macroscopic droplet, A is the parameter of its specific surface energy, and in the water case these parameters are equal to [48] $\varepsilon_o = 0.43\,\mathrm{eV}$, $A = 0.21\,\mathrm{eV}$ and relate to the liquid aggregate state of droplets at the temperature $T = 293\,\mathrm{K}$. This gives for the binding energy change $\Delta\varepsilon(n)$ if one molecule is added to a droplet consisting of n bound molecules

$$\Delta\varepsilon(n) = \frac{dE_n}{dn} - \varepsilon_o = \frac{2A}{3n^{1/3}} \tag{3.28}$$

The coalescence process is considered under an equilibrium between free atmospheric molecules of water and bound water molecules of large droplets, so that the criterion $A \ll T \cdot n^{1/3}$ holds true. In the first approximation under this equilibrium, the number density of free water molecules N_w is equal to that at the saturated vapor pressure $N_{sat}(T)$, i.e. $N_w = N_{sat}(T)$. Taking for the saturated vapor pressure $N_{sat}(T) \sim \exp(-\varepsilon_o/T)$, where ε_o is the molecule binding energy in bulk water, we obtain for the critical droplet size [35]

$$N_w = N_{sat}(T) \exp \left[\frac{\Delta \varepsilon(n_{cr})}{T} \right]. \tag{3.29}$$

One can represent this equation in the form

$$N_{sat}(T) - N_w = \left[\frac{2A}{3Tn_{cr}^{1/3}} \right] N_{sat}(T) \tag{3.30}$$

In particular, a small parameter which characterizes a deviation from the relation $N_w = N_{sat}(T)$ for water microdroplets of a cumulus cloud (3.1) is estimated as $\sim 10^{-4}$.

Accounting for the temperature dependence for the evaporation rate to be $\sim \exp(-\Delta \varepsilon(n)/T)$, one can obtain for the total flux of molecules to the droplet surface of a given size

$$J_t(n) = J_{at}(n) - J_{ev}(n) = 4\pi D_w r N_w \left\{ 1 - \exp \left[\frac{\Delta \varepsilon(n_{cr}) - \Delta \varepsilon(n)}{T} \right] \right\}, \tag{3.31}$$

where we use the Smoluchowski formula [76] for the number of molecules attaching to a droplet of a radius r per unit time. This formula demonstrates the competition of processes of molecule attachment and droplet evaporation in the total process of the droplet growth.

On the basis of the relation (3.27), one can construct the growth equation for a droplet of a given size n as

$$\frac{dn}{dt} = J(n) = \frac{8\pi D_w r_W N_w A}{3T} (u^{1/3} - 1), \tag{3.32}$$

where $u = n/n_{cr}$ is the reduced droplet size, a droplet size assumes to be large $n \gg (A/T)^3$, and the Wigner–Seitz radius r_W is introduced on the basis of formula (3.2). Because of the competition of the evaporation and attachment processes, an average of $u^{1/3} - 1$ gives a small value compared with one, namely [66]

$$\overline{u^{1/3}} - 1 = 0.056$$

From this, one can obtain the equation for the change of the average number $\bar{n}$ of droplet molecules as [66]

$$\frac{d\bar{n}}{dt} = J_a = 0.47 D_w r_W N_w \frac{A}{T} \tag{3.33}$$

As it follows from this, the average droplet radius r increases in time as $r \sim t^{1/3}$ as a result of the coalescence process.

Accounting for the thermal factor, the growth equation for the average number of droplet molecules $\bar{n} = 1.13 n_{cr}$ takes the form

$$\frac{d\bar{n}}{dt} = \frac{0.47 D_w r_W N_w A}{T \Phi(T)} = \frac{1}{\tau_c}, \quad \tau_c = \frac{2.1 \Phi}{D_w r_W N_{\text{sat}}} \frac{T}{A}, \tag{3.34}$$

where τ_c is characterized as an increase of the average number of droplet atoms by one. In particular, at the temperature $T = 273\,\text{K}$ and at atmospheric pressure, this parameter is equal to $\tau_c = 6.0 \times 10^{-10}\,\text{s}$.

From this, one can obtain for a time τ of droplet growth from an average molecule number $\bar{n}$ up to $2\bar{n}$ (or from zero up to this average number of droplet molecules $\bar{n}$ if this process proceeds in the diffusion growth regime), which is given by

$$\tau = \tau_c \bar{n} \tag{3.35}$$

Figure 3.12 gives the temperature dependence for the doubling time of liquid droplets in atmospheric air or a time of formation of these droplets in a cumulus cloud with parameters according to (3.1).

Fig. 3.11 Character of association of liquid droplets for the gravitation mechanism of growth of microdroplets

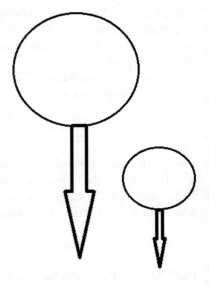

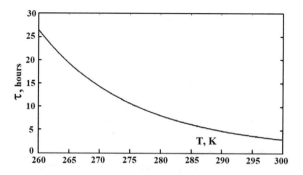

Fig. 3.12 Doubling time or a time of formation for an average number of water molecules in droplets of a cumulus cloud with parameters (3.1). The growth process in atmospheric air assumes to be in the diffusion regime of coalescence [66, 92]

3.3.2 Establishment of Equilibrium in Growth of Water Droplets

In this consideration, the equilibrium is supported between droplets and free water molecules of atmospheric air, and therefore the number density of free molecules is close to that at the saturated vapor pressure. The coalescence growth rate is proportional to the number density of free water molecules, whereas this rate is proportional to the number density of bound water molecules for the coagulation mechanism of droplet growth. Under these conditions, a typical time of establishment of this equilibrium τ_{eq} is small compared to a time (3.35) of droplet growth. We now consider the character of equilibrium if the number density of free molecules N_w differs from the equilibrium one N_{sat}, and the evolution of the droplet size leads simultaneously to the equilibrium $N_w = N_{sat}$.

The balance equation for a droplet size in the diffusion regime of these processes on the basis of the Smoluchowski formula (2.30) takes the form

$$\frac{dn}{dt} = 4\pi D_w r (N_w - N_{sat}) = \frac{n^{1/3}(N_w - N_{sat})}{\tau_{eq}}, \qquad (3.36)$$

where n is the number of bound molecules in a droplet, r is a current droplet radius, so that $n = (r/r_W)^3$, $r_W = 0.192$ nm is the Wigner–Seitz radius of a liquid water droplet, D_w is the diffusion coefficient of water molecules in air, N_w is the current number density of water molecules, and N_{sat} is the number density at the saturated vapor pressure. From this, one can define the relaxation time τ_{eq}

$$\tau_{eq} = \frac{1}{4\pi D_w r N_{sat}} \qquad (3.37)$$

In particular, under normal conditions, i.e. at the temperature $T = 273$ K and atmospheric pressure, this value in atmospheric air is equal to $\tau_{eq} \approx 3 \times 10^{-15}$ s. As is seen, a typical time of equilibrium establishment is several orders of magnitudes lower than a typical time of droplet growth.

Along with the equation for droplet growth, we have now the equation for the evolution of the number density of water molecules

$$\frac{dN_w}{dt} = 4\pi D_w r_o (N_{\text{sat}} - N_w).$$ (3.38)

This equation follows from the conservation of the total number density of water molecules $N_w + N_b$, where the number density of bound molecules in water droplets is equal to $N_b = N_d \bar{n}$, and the number density N_d of water droplets is unvaried during the relaxation process.

Let us represent equation (3.36) for a droplet size as a result of droplet evaporation and molecule attachment to its surface as

$$\frac{dr^2}{dt} = \alpha \left(\frac{N_w}{N_{\text{sat}}} - 1 \right), \quad \alpha = \frac{8\pi}{3\Phi(T)} r_W^3 D_w N_{\text{sat}}.$$ (3.39)

In particular, at the temperature $T = 273$ K and atmospheric pressure of air with parameters $D_w = 0.22 \, \text{cm}^2/\text{s}$ [77, 78], $N_{\text{sat}} = 1.6 \times 10^{17} \, \text{cm}^{-3}$, $\Phi_o(T) = 1.7$, we have $\alpha = 4 \times 10^{-7} \, \text{cm}^2/\text{s}$. From this, it follows that a water droplet located in a vacuum will be evaporated through a time $t_{ev} \approx 0.8$ s. As is seen, this time is small compared to typical times of droplet evolution. This means that the equilibrium between droplets and free water molecules is established fast, and the droplet growth proceeds under equilibrium conditions.

One can add to this that processes of droplet growth and evaporation are accompanied by heat extraction or release. As a result, in a stable regime of the growth or the evaporation process, the droplet temperature differs from the temperature of the surrounding air far from the droplet. One can use formula (2.35) for the temperature difference ΔT in this case as

$$\Delta T = \frac{\varepsilon_o D_w (N_w - N_{\text{sat}})}{\kappa \Phi(T)}$$ (3.40)

Let us consider the case of droplet evaporation in atmospheric air, where the moisture is small, i.e. $N_w \ll N_{\text{sat}}$. Under conditions which we use usually, we take air at atmospheric pressure and temperature $T = 273$ K. Let us take the binding energy of a bound water molecule $\varepsilon_o = 0.43$ eV, the diffusion coefficient of a water molecule in air $D_w = 0.22 \, \text{cm}^2/\text{s}$, the number density of water molecules $N_{\text{sat}} = 1.6 \times 10^{17} \, \text{cm}^{-3}$ at the saturated vapor pressure, and the thermal factor due to the feedback defined by formula (2.35) is equal $\Phi_o(T) = 1.7$ under these conditions. One can find finally that under these conditions, the droplet is cooled with respect to the surrounding air by the temperature $\Delta T = 6$ K.

3.3.3 Gravitation Mechanism of Growth of Water Droplets

The above analysis shows that the coalescence mechanism of droplet growth is the basic mechanism among other ones (Fig. 3.10) of growth at the initial growth stage. But this growth process decelerates strongly at large droplet sizes, and therefore it is replaced by the gravitation growth at large droplet sizes. The nature of the gravitation mechanism is represented in Fig. 3.11 and results in the merging of droplets which move in atmospheric air under the action of their weight. Let us start from an estimation the rate constant of gravitational association of droplets $k_{as} \sim \Delta v \cdot \sigma$, where Δv is the difference of velocities of two droplets, $\sigma = \pi(r_1 + r_2)^2$ is the cross section of contact between two droplets of radii r_1 and r_2. Assuming that each contact of two liquid droplets leads to their association and using formula (2.22) for the fall velocity of droplets, one can estimate the rate constant of association for droplets of a typical radius r

$$k_{as} \sim \frac{\rho g r^4}{\eta}.$$

The accurate expression for this rate constant has the form [65]

$$k_{as} = \frac{2\rho g r^4}{\eta}. \tag{3.41}$$

One can represent formula (3.41) in the form

$$k_{as} = \chi_o \left(\frac{r}{r_W}\right)^4 = \chi_o n^{4/3}, \quad \chi_o = \frac{2\rho g r_W^4}{\eta} \tag{3.42}$$

In particular, for water droplets in air at atmospheric pressure and temperature $T = 273\,\text{K}$ $(\eta = 1.45 \times 10^{-4}\,\text{g/(cms)})$, this parameter is equal to $\chi_o = 1.8 \times 10^{-16}\,\text{cm}^3/\text{s}$.

The gravitation mechanism of droplet association is stronger than the coagulation one, if the velocity of gravitation fall w_g exceeds significantly a thermal velocity of droplets, i.e. if the following criterion holds true

$$w_g \gg \sqrt{\frac{8T}{\pi M}}, \tag{3.43}$$

where M is the droplet mass, and T is the gas temperature expressed in energetic units. This criterion gives

$$r \gg \left(\frac{12\eta^2 T}{\rho^3 g^2}\right)^{1/7} \tag{3.44}$$

In particular, for water droplets located in atmospheric air at room temperature, this criterion takes the form

$$r \gg 2\,\mu m \tag{3.45}$$

We now consider the kinetics of droplet growth as a result of the gravitation mechanism of droplet association. In the case of the association of two droplets consisting of n_1 and n_2 molecules, the final number of droplet molecules becomes $n = n_1 + n_2$, whereas before the joining the average number of droplet molecules is $(n_1 + n_2)/2$, so that the average change of the number of droplet molecules is $n/2$. Correspondingly, we obtain the balance equation for the average number n of droplet molecules [67]

$$\frac{dn}{dt} = \frac{1}{2} k_{as} N_b$$

because N_b/n is the average number density of droplets, and N_b is the number density of bound molecules in droplets. We are guided by parameters of droplets of a cumulus cloud (3.1), where $r = 8\,\mu m$, $N_b = 7.2 \times 10^{16}\,cm^{-3}$. This balance equation may be represented in the form

$$\frac{dn}{dt} = \frac{n^{4/3}}{\tau_{gr}}, \quad \tau_{gr} = \frac{2}{\chi_o \cdot N_b} = 1.5 \times 10^7\,s \tag{3.46}$$

From this, one can find the total growth time t_{gr} from a given size up to infinite one as

$$t_{gr} = \tau_{gr} \int\limits_{n}^{\infty} \frac{dn}{n^{4/3}} = \frac{3\tau_{gr}}{n^{1/3}} \tag{3.47}$$

Under the considering conditions, this time equals to $t_{gr} = 18\,min$.

One can join mechanisms of coalescence and gravitation growth, and then the growth equation on the basis of formulas (3.35) and (3.46) takes the form

$$\frac{dn}{dt} = \frac{n^{4/3}}{\tau_{gr}} + \frac{1}{\tau_c} \tag{3.48}$$

From this, it follows the total growth time τ_t which is given by

$$\tau_t = \int\limits_{0}^{\infty} \frac{dn}{n^{4/3}/\tau_{gr} + 1/\tau_c} = 3.3 \tau_c^{1/4} \tau_{gr}^{3/4} \tag{3.49}$$

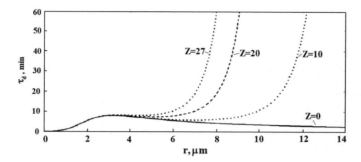

Fig. 3.13 Doubling time for the number molecules of water droplets according to formula (3.52). An indicated droplet charge Z is reached at a droplet radius $r = 8\,\mu m$. The temperature is $T = 273\,K$

Taking the values of times for coalescence ($\tau_c = 6.0 \times 10^{-10}\,s$) and gravitation growth $t_{gr} = 1.5 \times 10^7\,s$ according to formulas (3.34) and (3.46), one can obtain for the total growth time $\tau_t = 1.1\,h$.

Let us introduce the doubling time τ_d for the number of droplet molecules. During this time, a droplet radius varies from $2^{-1/6}r$ up to $2^{1/6}r$, i.e. the radius change Δr is

$$\Delta r = (2^{1/6} - 2^{-1/6})r = 0.23r$$

As a result, we have for the doubling time in this case

$$\tau_d(r) = \frac{0.7r^3}{r_W^3/\tau_c + r^4/(\tau_{gr} \cdot r_W)} \tag{3.50}$$

Figure 3.13 contains the doubling time of a microdroplet according to formula (3.50) for the combination of two growth mechanisms and under considering conditions. The maximum of the doubling time $\tau_d = 7.9\,min$ corresponds to a microdroplet radius of $3.1\,\mu m$. For the droplet radius (3.1) of an average cumulus cloud $r = 8\,\mu m$, the doubling time is equal to $\tau_d = 4.1\,min$.

We now consider the growth of charged droplets. The droplet charge does not act on the coagulation process because the droplet growth proceeds through free neutral molecules. On the contrary, in the case of gravitation growth a droplet charge prevents droplets from their approachment and joining. For simplicity, we assume droplets to have the same charge, and then formula (3.41) for droplet association takes the form [65]

$$k_{as} = \frac{2\rho g r^4 \exp\left(-\frac{Z^2 e^2}{2rT}\right)}{\eta}, \tag{3.51}$$

Table 3.2 Expressed in minutes, doubling times of growth of water droplets of a radius $r = 8\,\mu m$ and of a charge Z in atmospheric air at the temperature $T = 273\,K$ and atmospheric pressure. The doubling times are evaluated on the basis of formula (3.49) using growth parameters according to formulas (3.34) and (3.46)

Z, e	τ_d, min
0	4.1
2	4.2
5	4.6
10	6.1
15	9.7
20	19
25	42
30	104
40	400

where Z is the droplet charge which assumes below the same for all droplets, as well as their size. As a result, the doubling time $\tau_d(r)$ is equal for charged droplets instead of (3.50)

$$\tau_d(r) = \frac{0.7 r^3}{r_W^3/\tau_c + r^4 \exp\left(-\frac{Z^2 e^2}{2rT}\right)/(\tau_{gr} \cdot r_W)} \tag{3.52}$$

Here, we assume the droplet charges to be unvaried in combined droplets in the course of their joining. Figure 3.13 contains the doubling time of charged microdroplets in accordance with formula (3.52). We take the coalescence time $\tau_c = 6.0 \times 10^{-1}\,s$ according to formula (3.34), and $\tau_{gr} = 1.5 \times 10^7\,s$ according to formula (3.46). All the evaluations correspond to the air temperature $T = 273\,K$ and atmospheric pressure.

Table 3.2 contains the values of doubling times for charged droplets. For the growth mechanism under consideration, the specific droplet charge, i.e. the ratio of the charge to the droplet mass, does not vary in the course of droplet growth. Thus, in this analysis of droplet growth, we account for the coalescence and gravitation growth mechanisms in accordance with Fig. 3.10. From this, it follows that growth of charged droplets proceeds slowly at large droplet sizes, so that the total droplet growth time exceeds the doubling time for this microdroplet by one order of magnitude. In the case under consideration, where droplets of a cloud are charged with the same charge sign, the droplet growth is stopped at a certain size depending on the droplet charge.

Summing up the results of the kinetics of growth of water droplets in atmospheric air, we come to the following conclusions. Under real conditions, the equilibrium time between water vapor and the condensed phase of water in the atmosphere is small compared with a typical time of evolution of this system. From this it follows that the existence of the water condensed phase in atmospheric air is possible only in a supersaturated water vapor, because the average atmospheric humidity is below 100% at any altitude, i.e. condensation is possible only in restricted regions of the atmosphere due to fluctuations. As a result, the condensation process is an unequilibrium one and occurs under the action of vertical winds which transfer warm and wet air up in cold regions of the atmosphere. But even in atmospheric regions where the vapor condensation occurs, the number density of bound molecules which constitute droplets is less than the number density of free water molecules.

3.3.4 Displacement of Microdroplets of Cumulus Clouds

Clouds include the condensed phase of water and consist of water microdroplets or microparticles. This cloud structure, where condensed water is concentrated in some space regions, follows from the character of water condensation in the atmosphere. Clouds are not supported by forces which compel microparticles to remain in a restricted space regions, i.e. clouds are spread in time and this process determines the cloud lifetime. Our goal is to estimate lifetimes of clouds with respect to various mechanisms.

Representing a cumulus cloud as a system of bunch of water microdroplets in atmospheric air, we first determine a typical time of spreading of an individual bunch. Let us determine first a typical time of spreading of a microdroplet bunch as a result of the diffusion of microdroplets for the diffusion regime, where the mean free path of air molecules λ is small compared to the droplet radius r. Taking for air at atmospheric pressure $\lambda \sim 0.1\,\mu m$, one can see that the diffusion regime of motion is realized for microdroplets located in atmospheric air. The diffusion coefficient D_d of microdroplets in atmospheric air in this case is determined by formula [35, 67]

$$D_d = \frac{T}{6\pi r \eta},$$ (3.53)

where T is the air temperature $T = 273\,K$ expressed in energetic units, η is the air viscosity that equals $\eta = 2.1 \times 10^{-4}\,g/(cm\,s)$ at the temperature $T = 273\,K$. For the droplet radius $r_o = 8\,\mu m$, this formula gives $D_d = 1.2 \times 10^{-8}\,cm^2/s$ under normal conditions, then a droplet shifts at a distance approximately $0.2\,mm$ for a time of $1\,h$. This estimation shows that the diffusion motion of microdroplets of cumulus clouds in motionless air is negligible.

We also estimate the interaction of water microdroplets of clouds with winds if this interaction has a random character and proceeds mostly under the action of wind gusts. If the velocity of a microdroplet differs from that of the surrounding air by v, the friction force is given by the Stokes formula

$$F = 6\pi r \eta,$$ (3.54)

and this force tends to equalize a microdroplet velocity and that of surrounded air. Finally, this force compels microdroplets to move with a wind speed or leads to braking and stopping the air wind. The equation for microdroplet motion in an air jet and its solution has the form

$$m\frac{dv}{dt} = 6\pi r v \eta, \quad v = v(0)\exp(-t/\tau_{rel}), \quad \tau_{rel} = \frac{2r^2\rho}{9\eta}, \tag{3.55}$$

where v is the difference of velocities of a test microdroplet and the surrounding air, $m = 4\pi r^3 \rho/3$ is the droplet mass, and the diffusion regime of droplet motion is considered here. For $r = 8\,\mu m$, this formula gives $\tau_{rel} \approx 0.7\,ms$. From this on the basis of a typical wind velocity $v \sim 5\,m/s$, one can estimate a distance l which a microdroplet passes with respect to moving air $l \sim 0.4\,cm$. One can conclude that if cloud sizes exceed centimeters, displacement of their element is not important for the cloud stability.

The stability of a cumulus cloud determines the character of its existence and its formation. It is of importance that this cloud gathers droplets in a restricted region as a result of heat processes, rather than due to forces which attract droplets to this space region. The mechanism of cloud formation consists in a sharp decrease of the air temperature in a region where a nonsaturated water vapor is located. Because this system is not supported by some forces, there are various instabilities which decay this ensemble of particles. In this analysis, we are guided by a typical cumulus cloud with parameters (3.1). Then a simple way for the formation of this cloud results from the mixing of air layers from different altitudes. This takes place if an air flux from a near-surface layer rises up to altitudes 2–3 km. An inverse process leads to the decomposition of such a cloud, and according to above estimations (3.2.2), water droplets evaporate in a typical cumulus cloud at its heating by 6 K. This shows that air motion in the form of convection and streams is important for the formation and evolution of clouds. Principal cloud processes are impossible in motionless air.

It is clear that a cumulus cloud as a system consisting of air molecules and water microdroplets is spread in a space as a result of droplet diffusion. The above estimations show that this process is weak for typical cloud parameters. The basic mechanism of cloud decay is evaporation of microdroplets. If the air temperature does not vary in this process, due to an equilibrium between free and bound water molecules, the number density of free water is equal to that at saturated vapor pressure, and the total number density for both free and bound water molecules almost do not vary in this process. The basic mechanism of droplet growth at large droplet sizes results from their contact in the course of gravitation falling. This time is several minutes (see Fig. 3.13) for typical parameters of cumulus clouds if droplets are neutral, while observed lifetimes of cumulus clouds are several hours. From this, it follows that water microdroplets of cumulus clouds are charged. An average droplet charge is equal to 27 electron charges [66, 81] for typical parameters of a cumulus cloud and its indicated lifetime.

Thus, from the physical standpoint, a cumulus cloud is a complicated physical system whose formation and existence are supported by some real conditions and are governed by corresponding processes.

References

1. http://en.wikipedia.org/wiki/Cloud
2. https://pixabay.com/ru/photos
3. L. Howard, Philosoph. Mag. Ser. **1**(16), 62 (1802)
4. L. Howard, Philosoph. Mag. Ser. **1**(17), 65 (1803)
5. https://yandex.ru/search/?lr=213
6. https://c.pxhere.com/photos/9a/cc/photo-145978.jpg!d
7. https://get.pxhere.com/photo/cloud-sky-white-atmosphere-weather-storm-cumulus-australia-thunderstorm-dramatic-large-meteorological-phenomenon-cumulus-nimbus-atmosphere-of-earth-geological-phenomenon-781881
8. H.L. Green, W.R. Lane, *Particulate Clouds: Dust, Smokes and Mists* (Princeton, Van Nostrand, 1964)
9. H.R. Byers, *Elements of Cloud Physics* (University of Chicago, Chicago, 1965)
10. N.H. Fletcher, *The Physics of Rainclouds* (Cambridge University Press, London, 1969)
11. B.J. Mason, *Clouds, Rain and Rainmaking* (Cambridge University Press, Cambridge, 1975)
12. H. Pruppacher, J. Klett, *Microphysics of Clouds and Precipitation* (Reidel, London, 1978)
13. R.G. Fleagle, J.A. Businger, *Introduction to Atmospheric Physics* (Academic Press, San Diego, 1980)
14. R.R. Rogers, M.K. Yau, *A Short Course in Cloud Physics* (Pergamon Press, Oxford, 1989)
15. F.H. Ludlam, *Clouds and Storms: The Behavior and Effect of Water in the Atmosphere* (Penn State University Press, University Park, 1990)
16. K. Young, *Microphysical Processes in Clouds* (Oxford University Press, New York, 1993)
17. H.R. Pruppacher, J.D. Klett, *Microphysics of Clouds and Precipitation* (Kluwer, New York, 2004)
18. M. Satoh, *Atmospheric Circulation Dynamiccs and General Circulation Models* (Springer-Praxis, Chichester, 2004)
19. J. Straka, *Clouds and Precipitation Physics* (Cambridge University Press, Cambridge, 2009)
20. B.J. Mason, *The Physics of Clouds* (Oxford University Press, Oxford, 2010)
21. H. Pruppacher, J. Klett, *Microphysics of Clouds and Precipitation* (Springer, Dodrecht, 2010)
22. D. Lamb, J. Verlinde, *Physics and Chemistry of Clouds* (Cambridge University Press, Cambridge, 2011)
23. P.K. Wang, *Physics and Dynamics of Clouds and Precipitation* (Cambridge University Press, Cambridge, 2013)
24. A.P. Khain, M. Pinsky, *Physical Processes in Clouds and Cloud Modeling* (Cambridge University Press, Cambridge, 2013)
25. S. Twomey, *Atmospheric Aerosols* (Elsevier, Amsterdam, 1977)
26. P.C. Reist, *Introduction to Aerosol Science* (Macmillan Computer Publishing, New York, 1984)
27. W.C. Hinds, *Aerosol Technology: Properties, Behavior and Measurement of Airborne Particles* (Wiley, New York, 1999)
28. K. Friedlander, Smoke, Dust, and Haze. Fundamentals of Aerosol Dynamics (Oxford University Press, Oxford, 2000)
29. B.J. Mason, *The Physics of Clouds* (Claredon Press, Oxford, 1971)
30. J. Warner, Tellus **7**, 450 (1955)
31. W.R. Leaitch, G.A. Isaak, Atmosp. Environ. **25**, 601 (1991)
32. http://en.wikipedia.org/wiki/Liquid-water-content
33. E.P. Wigner, W.F. Seits, Phys. Rev. **46**, 509 (1934)
34. E.P. Wigner, Phys. Rev. **46**, 1002 (1934)
35. B.M. Smirnov, *Clusters and Small Particles in Gases and Plasmas* (Springer, NY, New York, 1999)
36. https://en.wikipedia.org/wiki/Moisture
37. https://en.wikipedia.org/wiki/Cloud-cover
38. R. Braham, J. Meteorol. **9**, 227 (1952)

39. O. Boucher, *Atmospheric Aerosols. Properties and Climate Impacts* (Springer, Dordrecht, 2015)
40. L.A. Matveev, *Clouds Dynamics* (Gidrometeoizdat, Leningrad, 1981; in Russian)
41. R. Houze, *Cloud Dynamics* (Academic Press, San Diego, 1993)
42. http://zebu.uoregon.edu/ph311/rb.gif
43. https://geography-revision.co.uk/gcse/weather-climate/atmospheric-circulation
44. https://en.wikipedia.org/wiki/Hadley-cell
45. https://en.wikipedia.org/wiki/Atmospheric-circulation
46. J. Huang, M.B. McElroy, J. Clim. **27**, 2656 (2014)
47. https://en.wikipedia.org/wiki/Dew-point
48. *Handbook of Chemistry and Physics*, ed. by D.R. Lide, 86 edn. (CRC Press, London, 2003–2004)
49. A. Wexler, J. Res. Nat. Bur. Stand. **80A**, 775 (1976)
50. *U.S. Standard Atmosphere* (Washington, U.S. Government Printing Office, 1976)
51. H. Stommel, J. Meteor. **4**, 91 (1947)
52. B.R. Taylor, M.B. Baker, J. Atmos. Sci. **48**, 112 (1991)
53. D. Gregory, Quart. J. Roy. Meteor. Sci. **127**, 153 (2001)
54. W.R. Cotton, R.A. Anthes, *Storm and Cloud Dynamics* (Academic Press, San Diego, 1989)
55. K.A. Emanuel, *Atmospheric Convection* (Oxford University Press, New York, 1991)
56. P. Squires, Tellus **10**, 256 (1958)
57. A.J. Heymsfield, P.N. Johnson, J.E. Dye, J. Atmos. Sci. **35**, 1689 (1978)
58. T. Heus, H.J.J. Jonker, J. Atmos. Sci. **65**, 1003 (2008)
59. T. Heus, J. Atmos. Sci. **65**, 2581 (2008)
60. YaB Zeldovich, ZhETF **12**, 525 (1942)
61. F.F. Abraham, *Homogeneous Nucleation Theory* (Academic Press, New York, 1974)
62. L.D. Landau, E.M. Lifshitz, *Statistical Physics*, vol. 1 (Pergamon Press, Oxford, 1980)
63. I. Gutzow, J. Schmelzer, *The Vitreous State* (Springer, Berlin, 1995)
64. B.M. Smirnov, *Principles of Statistical Physics* (Wiley VCH, Berlin, 2006)
65. B.M. Smirnov, *Nanoclusters and Microparticles in Gases and Vapors* (DeGruyter, Berlin, 2012)
66. B.M. Smirnov, Phys. Usp. **57**, 1041 (2014)
67. B.M. Smirnov, *Cluster Processes in Gases and Plasmas* (Wiley, Berlin, 2010)
68. M.M.R. Williams, S.K. Loyalka, *Aerosol Science. Theory and Practice* (Pergamon, Oxford, 1991)
69. A.A. Lushnikov, *Introduction to aerosols*, in *Aerosols–Science and Technology*, ed. by I. Agranovski (Wiley, Weinheim, 2010), pp. 1–42
70. A.A. Lushnikov, *Nanoaerosols in the atmosphere*, in *The Atmosphere and Ionosphere, Physics of Earth and Space Environments*, ed. by V.L. Bychkov et al. (Springer, Dordrecht, 2012), pp. 79–164
71. A.A. Lushnikov, V.A. Zagaynov, Y.S. Lyubovtseva, Formation of aerosols in the atmosphere, in *The Atmosphere and Ionosphere, Physics of Earth and Space Environments*, ed. by V.L. Bychkov et al. (Springer Science, Dordrecht, 2012), pp. 69–95
72. A.A. Lushnikov, Condensation, evaporation, nucleation, in *Aerosols–Science and Technology*, ed. by I. Agranovski (Wiley, Weinheim, 2010), pp. 91–126
73. N.A. Fuchs, *Evaporation and Growth of Drops in a Gas* (Izd. AN SSSR, Moscow, 1958; in Russian)
74. N.A. Fuchs, A.G. Sutugin, *Highly Dispersed Aerosols* (Ann Arbor, London, 1971)
75. N.A. Fuchs, *Mechanics of Aerosols* (Pergamon, New York, 2002)
76. M.V. Smolukhowski, Zs. Phys. **17**, 585 (1916)
77. B.M. Smirnov, *Reference Data on Atomic Physics and Atomic Processes* (Springer, Heidelberg, 2008)
78. N.B. Vargaftic, *Tables of Thermophysical Properties of Liquids and Gases* (Halsted Press, New York, 1975)
79. J.H. Seinfeld, S.N. Pandis, *Atmospheric Chemistry and Physics* (Wiley, Hoboken, 2006)

80. M.L. Salby, *Physics of the Atmosphere and Climate* (Cambridge University Press, Cambridge, 2012)
81. B.M. Smirnov, *Microphysics of Atmospheric Phenomena* (Springer Atmospheric Series, Switzerland, 2017)
82. K.S.W. Champion, A.E. Cole, A.J. Kantor, in *Chemical Dynamics in Extreme Environments*, ed. by R.A. Dressler (World Scientific Publishing, Singapore, 2001)
83. B. Sportisse, *Fundamentals in Air Pollution* (Springer Science, Dordrecht, 2010)
84. W. Ostwald, Zs. Phys. Chem. **22**, 289 (1897)
85. W. Ostwald, Zs. Phys. Chem. **34**, 495 (1900)
86. I.M. Lifshitz, V.V. Slezov, JETP **35**, 331 (1958)
87. I.M. Lifshitz, V.V. Slezov, JETP **1**, 1401 (1959)
88. I.M. Lifshitz, V.V. Slezov, J. Phys. Chem. Sol. **19**, 35 (1961)
89. E.M. Lifshitz, L.P. Pitaevskii, *Physical Kinetics* (Pergamon Press, Oxford, 1981)
90. V.V. Slezov, V.V. Sagalovitch, Sov. Phys. Usp. **30**, 23 (1987)
91. S. Ino, J. Phys. Soc. Jpn. **27**, 941 (1969)
92. B.M. Smirnov, EPL **99**, 13001 (2012)

Chapter 4
Processes Involving Atmospheric Ions

Abstract Processes of ion formation in the atmosphere and their decomposition, as well as the behavior of ions in the troposphere, are analyzed. Cosmic rays including protons and neutrons of GeV energies, as well as hard radiation of galactic or solar origin, penetrate in the atmosphere and lead to the ionization of air molecules through nuclear reactions. The formation of ions under the action of cosmic rays proceeds in all of the troposphere with the rate of the order of $10\,\mathrm{cm}^{-3}\mathrm{s}^{-1}$. Charging of water microdroplets and microparticles which form clouds takes place both in pair collisions if the colliding particles which are located in different aggregate states and due to the attachment of atmospheric ions to microdroplets owing to different mobilities of positive and negative ions. Processes of ion recombination in atmospheric air and transport of tropospheric ions are analyzed. As a result of atmospheric processes, the Earth is charged negatively. Because of a large lifetime of atmospheric ions with respect to recombination, they are covered by a coat consisting of water and other molecules. Since the mobility of ions decreases with an increase of their size, during basic time of residence in the atmosphere growing ions do not partake in the electric processes. We divide the atmosphere into two parts, a clear-sky atmosphere, where discharging of the Earth occurs, and cumulus clouds which include a small part of the atmosphere with the basic portion of atmospheric condensed water in the form of microdroplets mostly. According to an equilibrium between free water molecules and water microdroplets in cumulus clouds, the number density of free water molecules is equal to that of the saturated one at the temperature of a cloud. Gravitation falling of charged water microdroplets of cumulus clouds provides separation of charges in the atmosphere and is responsible for the Earth's charging.

B. M. Smirnov, *Global Atmospheric Phenomena Involving Water*,
Springer Atmospheric Sciences, https://doi.org/10.1007/978-3-030-58039-1_4

4.1 Cosmic Rays in Atmospheric Ionization

4.1.1 Cosmic Rays in the Earth's Atmosphere

Cosmic rays, mostly protons and neutrons of high energies, which penetrate in the Earth's atmosphere, are responsible for atmospheric ionization. The history of study of cosmic rays as a source of atmospheric ionization involves a number of stages, and usually as its beginning is taken the measurement by Hess [1], a scientist and pilot, who in 1912 raised the electroscope on a balloon up to a height of 5 km in order to study the tropospheric ionization in a series of experiments. Hess's measurements led to a contradiction because in that time the atmospheric ionization was explained by the radioactivity of the Earth. Under this assumption, the rate of ionization must decrease as the altitude increases, whereas Hess's measurements led to another altitude dependence for the ionization rate, namely, starting from a certain altitude, the rate of ionization of atmospheric air increases with altitude.

Until that time, measurements were made for ionization of the near-surface atmosphere, which was carried out in the early twentieth century by different methods. In particular, results of such studies were given in publications [2–8]. At low altitudes, the ionization of atmospheric air studied in these experiments was associated with soil radioactivity, and this concept was confirmed in corresponding measurements. For example, according to the experiment by Mclennon and Barton [9], 5 cm of lead reduced the rate of ionization in the atmosphere by 30%, which indicated that particles of high energies from a soil cause air ionization. Subsequent studies have shown that the soil radioactivity is responsible for the ionization of air at low altitudes. An important stage of such studies were measurements of the ionization rate of the atmosphere at the Eiffel Tower by T. Wulf on the basis of his improved electroscope [10]. It was found that the rate of ionization of the atmosphere near the Earth's surface is twice compared to the top of the Eiffel Tower at an altitude of 300 m. This shows that the ionization rate due to the Earth's radioactivity at large altitudes must be much lesser.

A series of Hess measurements [1] becomes an important stage of the discovery of cosmic rays. Subsequently, Wilson [11] on the basis of a modified Wilson chamber for ion registration in balloon experiments exhibited that day and night measurements did not lead to a fundamental difference in the rate of ionization. A great contribution to the study of cosmic rays was made by Mulliken [12], who introduced the term "cosmic rays" and demonstrated in balloon experiments that the growth of the ionization rate for the atmosphere under the action of cosmic rays decreases at altitudes above approximately 10 km. In addition, ionization under the action of cosmic rays occurs not only in the atmosphere, but also in water and lakes, where the effect of the radioactivity of the Earth is excluded that testified about an extraterrestrial nature of cosmic rays. In addition, it was found that the ionization rate was identical at points with the same geographical latitude.

Fig. 4.1 Magnetic lines of force for the Earth

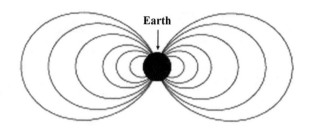

Mulliken considered cosmic rays first as high-energy photons (γ-rays) [12] of cosmic origin, and subsequently, the concept of cosmic rays was expanded to elementary particles of high energies [13, 14]. At this stage of the study of cosmic rays in the atmosphere, some elementary particles were discovered, such as π-mesons and μ-mesons. At the next stage of this research, accelerators were constructed for this goal, and another character of study of elementary particles as cosmic rays proceeded [15–22].

4.1.2 Excitation of Atmospheric Air by Fast Protons

Taking fast protons and neutrons as the basic components of cosmic rays, we note that according to Fig. 4.1 only fast protons and neutrons may penetrate in the Earth's atmosphere. Indeed, the Earth is a magnetic dipole, and the possibility for particles with a magnetic moment can overtake the magnetic field. The characteristic of this process is the magnetic cutoff rigidity R_c [23–25], which is proportional to the magnetic moment of an incident particle. In particular, for protons, as the most common nucleon component of cosmic radiation, the magnetic cutoff rigidity is $R_c = 14\,\text{GeV}$ [26]. Protons can reach the Earth's surface at the equator if its energy exceeds R_c, whereas slow protons may attain the Earth's surface at poles. Thus, the Earth acts with its magnetic field on nucleons of cosmic rays as a mass spectrometer [27–30].

We give in Fig. 4.2 the spectrum of cosmic rays which propagate at the Earth's level in a cosmos. One can see that cosmic rays which penetrate in the Earth's atmosphere and may cause its ionization are available both for measurements and generation in a contemporary setup. Considering cosmic rays from the modern point of view, we note that the low-energy part of their spectrum is of interest for ionization processes in the atmosphere. For comparison, this figure shows the energy achieved at the CERN Collider in 2015 (6.5 TeV).

Let us consider now the braking of fast protons passed through the atmosphere under the following conditions. In this consideration, we take into account the scattering of a fast proton on each bound atmospheric electron as a result of the Coulomb interaction between them in a nonrelativistic case where the proton velocity is small compared to the light speed. Along with this, we assume the proton velocity to be

Fig. 4.2 Low-energy part of
the cosmic ray spectrum
[31–33]. The arrow indicates
the maximum proton
collision energy reached at
the CERN Collider [34, 35]

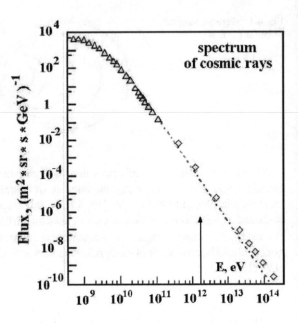

large compared to a typical electron velocity which moves along its orbit. Then one
can use the Landau–Lifshitz theory [37] with an independent scattering of various
electrons.

Within the framework of the Landau–Lifshitz theory, the variation of the energy
E of a fast proton is equal to

$$\frac{dE}{dz} = -N_e\kappa = -N_e \int \Delta\varepsilon d\sigma(\Delta\varepsilon), \qquad (4.1)$$

where z is the direction of motion, N_e is the number density of bound electrons in
air molecules, and $\sigma(\Delta\varepsilon)$ is the differential cross section of particle scattering on
a bound electron with the energy transfer $\Delta\varepsilon$. Summarizing over excited states of
air molecules, with accounting for states of continuous spectrum for electrons, one
can obtain for the specific braking force κ of a fast proton in a gas with taking into
account the sum rule for the oscillator strengths of electron transitions. This gives
the following expression [37]

$$\kappa = \frac{4\pi e^4 \ln \Lambda}{m_e v^2} = \frac{2\pi e^4 \ln \Lambda}{E} \cdot \frac{M}{m_e}, \quad v \ll c \qquad (4.2)$$

where e is the charge of an electron and of the incident particle (proton), M, v, and
E are the mass, velocity, and energy of the incident particle (proton), respectively,
m_e is the electron mass, c is the light speed, and $\ln \Lambda$ is the Coulomb logarithm.

We also take into account in formula (4.2) the coincidence of the cross sections
in quantum and classical cases for the Coulomb interaction of colliding particles. In

this model, the interaction of the proton with each electron is weak, and therefore the total scattering amplitude is the sum of those for each electron. Next, the Coulomb logarithm is equal to

$$\ln \Lambda = \ln \left(\frac{v}{v_o} \right)^2 \tag{4.3}$$

where v_o is a typical electron velocity in its orbits. In particular, within the framework of the Thomas–Fermi model for bound electrons, we have for their typical velocity $v_o \sim e^2 Z^{1/3}/\hbar$, where Z is the core charge of an air molecule. Taking $Z = 7$ for air molecules and $E = 10\,\text{MeV}$ for the proton energy, we have $\ln \Lambda \approx 5$. Though the Coulomb logarithm depends weakly on collision parameters, one can use this value for other cases. The number density N_e of electrons is connected to the number density of air molecules N_m as

$$N_e = 2Z N_m, \tag{4.4}$$

where $Z \approx 7$ is the average charge of nuclei.

Thus, from (4.1) for a current proton energy $E(z)$ as a function of an altitude z counted from some origin, we have

$$\frac{\mathrm{d}E}{\mathrm{d}z} = \frac{4\pi Z e^4 N_a(z) \ln \Lambda}{E} \cdot \frac{M}{m_e} \tag{4.5}$$

If we take the standard dependence of the number density of air molecules $N(z)$ on the altitude z as $N_m(z) \sim \exp(-z/\Lambda)$, one can obtain the dependence of an altitude h from the Earth's surface, where an incident proton stops on the initial energy e on the basis of (4.5)

$$E^2(h) = \pi Z e^4 L N_m(h) \ln \Lambda \cdot \frac{M}{m_e} \tag{4.6}$$

Figure 4.3 gives this dependency.

Note that in this analysis, basing on 1he Landau–Lifshitz theory [37], we neglect elastic scattering of an incident proton on nuclei of molecules. Let us check the validity of this assumption. We have for the mean free path λ_{el} of an incident proton with respect to elastic scattering on molecular nuclei due to Coulomb interaction

$$\lambda_{el} = \frac{1}{2N_m \sigma}, \quad \sigma = \frac{\pi Z^2 e^4}{E^2} \ln \Lambda, \tag{4.7}$$

where σ is the cross section of elastic scattering for the Coulomb interaction between an incident proton and molecular nuclei, Z is the nuclear charge, and the factor 2 takes into account two nuclei per molecule. Comparing this with the above mean free path λ_{in} for the energy loss due to inelastic scattering of a proton on bound electrons, one can obtain

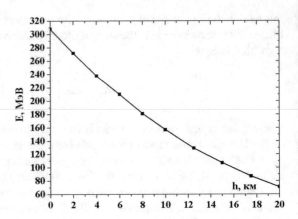

Fig. 4.3 Dependence of the initial proton energy E on an altitude h where a proton stops if its in motion in the Earth's atmosphere is directed perpendicular to its surface [36]

$$\frac{\lambda_{el}}{\lambda_{in}} = \frac{2M}{Zm_e} \gg 1 \tag{4.8}$$

From this, it follows that the energy loss of a fast proton, when it is brakes in a molecular matter, is determined by inelastic processes, i.e. by processes of excitation and ionization of molecules.

As it follows from Fig. 4.3, protons with an energy about 1 GeV and more are not braked in air as a result of the interaction of incident protons with electrons of air molecules. On the other hand, the Earth's magnetic field does not allow for protons and neutrons of such an energy to penetrate the Earth's atmosphere. From this, one can conclude that another character of interaction is responsible for the ionization of atmospheric air by cosmic rays.

4.1.3 Character of Absorption of Cosmic Rays by Atmosphere

Let us consider the process of interaction of an incident particle with air molecules from another standpoint. Let us introduce the cross section σ of this interaction and determine the distribution for altitudes at which this process proceeds. Evidently, a typical altitude h where the ionization process proceeds is given by

$$\Lambda_N \sigma N_m \sim 1 \tag{4.9}$$

Here, $\Lambda_N \approx 8$ km is the typical thickness of the atmosphere, and $N_m \sim 10^{19}$ cm^{-3} is a typical number density of air molecules.

For a more accurate determination of this quantity, let us introduce the intensity $I(h)$ of incident photons which move in the perpendicular direction to the Earth's surface. The equation for this intensity at an altitude h has the form

$$\frac{dI(h)}{dh} = -I\sigma N(h),\qquad(4.10)$$

where $N(h)$ is the number density of air molecules at an altitude h which is approximated by an exponential dependence in accordance with (2.3)

$$N(h) = N_o \exp\left(-\frac{h}{\Lambda}\right),\qquad(4.11)$$

and the scale length Λ is given as usual

$$\Lambda = \left(-\frac{d\ln N(h)}{dh}\right)^{-1}\qquad(4.12)$$

The solution of this equation for the surviving probability $P(h) = I(h)/I_o$, where I_o is the intensity of an incident flux, has the form

$$\ln\left[\frac{I(h)}{I_o}\right] = -\Lambda\sigma N(h),\qquad(4.13)$$

Let us introduce the parameter h_o as

$$\Lambda\sigma N(h_o) = \ln 2,\qquad(4.14)$$

and the variable $x = (h - h_o)/\Lambda$. Then the surviving probability $P(x)$ as a function of this variable has the form

$$P(x) = \exp\left[-\ln 2 \cdot \exp(-x)\right]\qquad(4.15)$$

Correspondingly, its derivative is given by

$$p(x) = \frac{dP(x)}{dx} = P(x)\ln 2\exp(-x)\qquad(4.16)$$

Dependencies $P(x)$ and $p(x)$ are represented in Fig. 4.4.

We now analyze from this standpoint the observed dependence of the ionization rate of the atmosphere by cosmic rays that is given in Fig. 4.5. The maximum ionization rate of atmospheric air is observed at altitudes of 11–15 km (so-called Pfotzer maximum) [40]) and is 30–40 cm^{-3}s^{-1} [39, 41, 42]. In addition, the total ionization rate of air per unit surface area of the Earth is 4.5×10^7 cm^{-2}s^{-1} [43]. The ratio of these values gives a typical range of altitudes $h \sim 10$ km, which gives the main contribution to atmospheric ionization. Since in this altitude range a typical number density of air molecules is $N \sim (10^{18}$–$10^{19})$ cm^{-3}, from this one can estimate a typical cross section of ionization of air molecules by cosmic rays as $(10^{-25}$–$10^{-24})$ cm^2.

Fig. 4.4 Dependence on the reduced altitude $x = (h - h_o)/\Lambda$ of the surviving probability $P(x)$ and its derivative $p(x)$ [36]

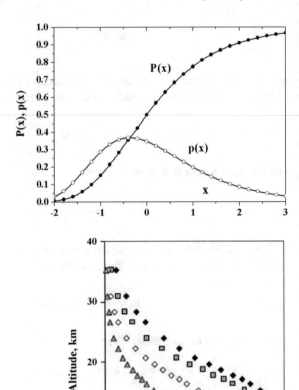

Fig. 4.5 Ionization rate of atmospheric air by cosmic rays at different altitudes [38, 39]

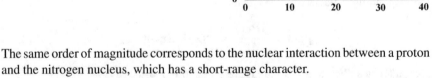

The same order of magnitude corresponds to the nuclear interaction between a proton and the nitrogen nucleus, which has a short-range character.

As for the near-surface regions of the atmosphere, a certain contribution to their ionization is given by the Earth radioactivity [3, 4], and therefore the rate of ionization in the near-surface atmosphere depends on the meteorological conditions [8]. A typical rate of air ionization in the near-surface regions of the atmosphere is (4–8) $cm^{-3}s^{-1}$ [44], although near the Earth's surface with increased radioactivity the rate of air ionization may reach the value of $10 \, cm^{-3}s^{-1}$ [45]. The rate of ionization of air near the ocean surface, where ionization is determined only by cosmic radiation, is equal to approximately $2 \, cm^{-3}s^{-1}$ [46]. The ionization of the air in near-surface regions of the atmosphere over land is also determined by γ-rays and α-particles created by soil and stones, so that the ionization rate of the air reaches $\sim 4 \, cm^{-3}s^{-1}$ at altitudes 200–300 m [47], and α-particles there result from the decay of radon nuclei Rn_{222}. Thus, soil radioactivity is important for air ionization in the lower layers of the atmosphere.

On the basis of data of Fig. 4.5, one can find a more accurate value of the cross section of the nuclear reaction with fast protons or neutrons which overcome the Earth's magnetic field and penetrate in the atmosphere. According to this figure, the maximum ionization due to these particles takes place at altitudes 11–15 km. Assuming that these altitudes correspond also to the maximum of the rate of the nuclear reaction, one can find the cross section σ of the nuclear reaction on the basis of formula (4.14)

$$\sigma = (1.1 \pm 0.5) \times 10^{-25} \, cm^2 \tag{4.17}$$

This corresponds to a typical value of such nuclear reactions and testifies about their short-range interactions in the course of nuclear reactions. One can find some analogy of this process with that studied in nuclear physics. In particular, the cross section for the process

$$n +^{14} N \rightarrow p +^{14} C \tag{4.18}$$

is $1.3 \times 10^{-25} \, cm^2$ at the collision energy of 13 MeV [48]. This corresponds to the result (4.17) because of a short-range character of interaction of colliding nucleons that is in agreement with the character of nuclear reactions [49, 50] and general principles of nuclear physics [51–55]). But in the case of collisions with GeV energies, the nuclear process becomes more rich because new channels are possible and the collision process has a relativistic character. Namely, due to a large collision energy, showers of elementary particles (as π-mesons) may be formed as a result of proton interaction with nitrogen or oxygen nuclei. Then an energy of an incident proton or neutron is extracted in a small space region and finally is consumed on in the ionization of molecules.

Note that if we consider the nuclear reaction involving a proton or neutron and the nitrogen or oxygen nucleus, on the basis of the short-range character of interaction, one can represent the cross section of the nuclear process as

$$\sigma = \pi R_n^2, \tag{4.19}$$

where the radius of the nitrogen or oxygen nucleon is $R_n \approx 2 \times 10^{-13}$ cm and exceeds significantly that for the proton or neutron. Hence, we obtain $\sigma \approx 10^{-25} \, cm^2$, and this estimation is in agreement with formula (4.17). Thus, the first stage of braking of fast protons and neutrons in the atmosphere results in the nuclear process involving nitrogen and oxygen nuclei. This nuclear reaction has the cascading character [21, 54], so that the energy of incident protons or neutrons is divided between several forming particles, which then interact more efficiently with air molecules because their energies are in the keV scale. As a result, the secondary particles of the nuclear process ensure the transfer of their energy to ionization of atmospheric air.

4.1.4 Photoionization of Air Molecules in the Atmosphere

In analyzing the atmospheric ionization by cosmic rays, we consider ionization processes as a result of penetration of fast protons and neutrons as the basic corpuscular component of cosmic rays in the atmosphere. But along with this, cosmic rays contain γ-radiation or short-wave photons which also interact weakly with air molecules and therefore can penetrate in the troposphere [56, 57]. Below, we consider the photoionization of air molecules by hard radiation which can penetrate in the Earth's troposphere. We are guided by such photons for which the cross section of the photoionization of air molecules ranges $(10^{-24}-10^{-25})$ cm^2 and hence such photons can reach altitudes of 11–15 km where the maximum atmospheric ionization is observed. Evidently, the energy of such photons greatly exceeds the binding energy of electrons from the K-shell of nitrogen and oxygen atoms. In this energy region, the ionization cross section decreases monotonically with an increasing photon energy. Therefore, in order to obtain the maximum ionization through this channel at altitudes of 11–15 km, it is required that the radiation source has a resonant nature, i.e. contained photons in a narrow energy range.

Being guided by the Sun as the basic source of cosmic rays, it is necessary to take into account the periodicity of the solar activity The basic period in solar activity oscillations is about 11 years, as it was discovered by Herschel in the early nineteenth century [58]. As it follows from Fig. 4.6, oscillations in the intensity of cosmic rays are observed with an indicated period. This means that the Sun is one of sources of cosmic rays which cause atmospheric ionization. Our task now is to ascertain what the role of hard radiation is in this process. In addition, if solar hard radiation gives a contribution to the ionization of the atmosphere at altitudes 11–15 km, it is necessary to establish the resonant character of this radiation.

Evidently, the solar corona [59–61] may be a source of hard radiation. The corona temperature in a quiet period is $(1-2) \times 10^6$ K, and the electron density in the corona plasma is of the order of 10^9 cm^{-3}. These values of the temperature and electron number density increase by several orders of magnitude during solar flares [56, 57], when large masses of the Sun erupt from the inner part of the Sun and interact with

Fig. 4.6 Evolution of the ionization rate of atmospheric air by cosmic rays at different altitudes [38, 39]

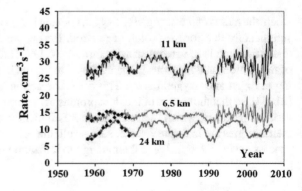

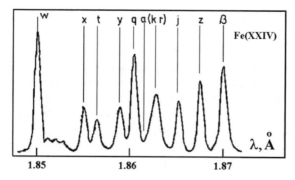

Fig. 4.7 Spectra of irons ion contained two bound electrons which are observed in the emission of the Sun's corona during a solar flare [62–64]. This spectrum is described by the electron temperature $T_e = 2.5 \times 10^7$ eV, which more or less correspond to the observed relative intensities of the corresponding satellites for the solar plasma

the solar corona. The propagation of the solar corona plasma into the surrounding space takes place in the form of the solar wind containing electrons and protons. In addition, the solar corona is a source of hard radiation with a wavelength of several Å. This radiation is formed in processes involving multicharge ions as a result of their photorecombination and dielectric recombination (for example, [62, 63]).

Evidently, the most likely source of solar hard radiation which can cause ionization of the Earth's atmosphere at the boundary of the troposphere and stratosphere (11–15 km over the Earth's surface), results from the radiative transition of the two-electron iron ion $FeXXIV(1s2p \longrightarrow 1s^2)$ with a wavelength in the range $\lambda = (1.85–1.87)$ Å. Figure 4.7 represents an example of a measured radiative spectrum of the solar corona at a certain electron temperature of the solar plasma. Emitted photons result from the radiative transition of an inner electron between $2p$ and $2s$ electron states, and satellite spectral lines are shifted with respect to the principal spectral line. Note that radiative transitions due to the iron multicharge ion give the main contribution to the solar radiation in this spectrum range.

Let us consider this concept from another standpoint and determine the cross section of the photo ionization process. In the case where the photon energy $\hbar\omega$ exceeds significantly the ionization potential J or a binding energy of a releasing electron, the asymptotic expression for the photoionization cross section for a hydrogen-like ion with the nucleus charge Z has the form [37, 65]

$$\sigma_\omega = \frac{\sigma_o}{Z^2} \cdot \left(\frac{J}{\hbar\omega} \right)^{7/2} \tag{4.20}$$

Here, $\sigma_o = 1.1$ Å^2. Below, we take into account that the K-electron shell of air molecules contains two electrons, and the transition of each of them contributes to the photoionization section of air molecules. Note that in reality, the above asymptotic expression (4.20) for the photoionization section holds true under the condition $\hbar\omega \gg$

$2\pi^2 J$ that is stronger than $\hbar\omega \gg J$ where formula (4.20) may be considered as an estimation.

We now use formula (4.20) for atmospheric air which consists mostly of nitrogen molecules. Hard radiation is emitted by iron multicharge ions of the solar corona with a wavelength of $\lambda = 1.86$ Å, which corresponds to the energy of the photon $\hbar\omega = 6.67$ Kev. Since $J = 667$ eV for K-electron shell of a nitrogen ion with $Z = 7$, formula (4.20) gives for the photoionization section of the nitrogen molecule $\sigma_\omega = 7.1 \times 10^{-22}$ cm^2 at the indicated wavelength. Taking $\Lambda = 8$ km, we obtain according to the formula (4.9) for the number density of nitrogen molecules at an altitude h_o, where the maximum absorption is realized

$$N_m(h_o) = \frac{\ln 2}{\sigma_\omega \Lambda_N} = 1.6 \times 10^{15} \, \text{cm}^{-3} \tag{4.21}$$

This gives the number density of nitrogen molecules $N_m(h_o) = 8 \times 10^{14}$ cm^{-3} that for the standard atmosphere model [66] corresponds to the altitude $h_o = 75$ km. The cross section of photoionization for the oxygen molecule under the action of resonant radiation is $\sigma_\omega = 1.4 \times 10^{-21}$ cm^2 ($Z = 8$, $J = 871$ eV). Accounting for the presence of oxygen molecules in the atmosphere, one can find the number density of air molecules $N_m(h_o) = 7 \times 10^{14}$ cm^{-3} for the optimal absorption at a given wavelength, which leads to the altitude $h_o = 76$ km with the maximum absorption. Such a large difference between observation and estimated altitudes proves that this character of atmospheric ionization through hard radiation from the solar corona is not realized.

Note in the case of radiative atmospheric ionization, the maximum altitude $h_o = 11$ km is realized according to formula (4.20), if the photon energy of incident γ-radiation is equal to $\hbar\omega = 93$ Kev, which refers to the wavelength $\lambda = 0.13$ Å. If this resonant photon occurs as a result of the radiative transition $2p \to 1s$ for the K-electron shell of an atomic particle, this corresponds to the nucleus charge $Z = 96$. As is seen, it is unrealistic. Thus, from this analysis one can conclude that γ-radiation is not responsible for the ionization of the troposphere.

4.2   Charging Processes in Low Atmosphere

4.2.1   Droplet Charging in Plasma

As a result of the action of atmospheric electricity, the Earth is charged negatively, and is discharged by an electric current which is created by ions formed by cosmic rays. In addition, lightning is the channel for the Earth's charging [67]. The latter proceeds from a breakdown between clouds and the Earth, and hence the negative charge to the Earth is transferred from clouds. This requires the separation of positive and negative charges of the atmosphere, and the only mechanism which can provide the separation of charges in the atmosphere is the gravitation falling of negatively charged microdroplets or microparticles. This means that atmospheric microdroplets are charged negatively, and we consider below the mechanisms which may provide this.

Let us consider charging of a microdroplet located in weakly ionized air, if a droplet radius greatly exceeds the mean free path of molecules in air. This corresponds to the diffusion regime of ion attachment to a droplet [68], and we derive expressions for ion currents to the droplet surface [69, 70]. In this regime, positive and negative ions attach to a droplet as a result of their diffusion motion in air and transfer their charge to the microdroplet. Let us take for definiteness, the droplet charge to be equal to $-Ze$ and assume that each contact of an ion with the microdroplet leads to ion attachment, i.e. the number density of ions at the droplet surface is $N_+(r) = 0$, where $N_+(R)$ is the number density of ions at a distance R from it. Considering the stationary problem, where the ion number density $N_+(R \to \infty) = N_o$ is supported far from the droplet, one can obtain the following expression for the current of positive ions I_+ toward the droplet surface as

$$I_+ = 4\pi R^2 \left(-D_+ \frac{dN_+}{dR} + K_+ E N_+ \right) e, \tag{4.22}$$

where D_+ is the diffusion coefficient of ions in air, K_+ is the mobility of ions, and E is the electric field strength which assumes to be relatively small.

The first term of the right-hand side of this equation describes the diffusion motion of ions, and the second term corresponds to the drift motion. Taking the electric field strength of the droplet as $E = Ze/R^2$, one can reduce this equation on the basis of the Einstein relation [71–73] $D_+ = eK_+/T$, where T is the gaseous temperature, to the form

$$I = -4\pi R^2 D_+ e \left(\frac{dN}{dR} - \frac{Ze^2 N}{TR^2} \right) \tag{4.23}$$

The solution of this equation under the above boundary condition $N_+(r) = 0$ has the form

$$N_+(R) = \frac{I_+}{4\pi D_+ e} \int_r^R \frac{dR'}{(R')^2} \exp\left(\frac{Ze^2}{TR'} - \frac{Ze^2}{TR} \right) = \frac{IT}{4\pi D_+ Ze^3} \left[\exp\left(\frac{Ze^2}{Tr} - \frac{Ze^2}{TR} \right) - 1 \right] \tag{4.24}$$

On the basis of another boundary condition $N_+(R \to \infty) = N_o$, one can obtain the relation between the ion current and the number density of ions far from the droplet as

$$I_+ = \frac{4\pi D_+ N_+ Ze^3}{T\{1 - \exp[-Ze^2/(Tr)]\}} \tag{4.25}$$

This formula is called the Fuks formula [74]. In the limiting case of a small charge $Ze^2/r_o \ll T$, this formula transfers into the Smoluchowski formula (2.30), and in the limit $Ze^2/r \gg T$, this formula is transformed into the Langevin formula [75]

$$I_+ = ZeN_+k_{as}, \quad k_{as} = 4\pi ZeK_+, \tag{4.26}$$

and K_{as} is the association rate constant of two charged particles of a different sign which are located in a dense buffer gas.

By analogy with formula (4.25), one can obtain an expression for the current of negative ions by changing $Z \to -Z$. We have

$$I_- = \frac{4\pi D_- N_- Ze^3}{T \cdot \left[1 - \exp\left(-\frac{Ze^2}{Tr}\right)\right]} \tag{4.27}$$

In the stationary regime, we have $I_+ = I_-$ that gives for the droplet charge

$$Z = \frac{rT}{e^2} \ln\left(\frac{K_- N_-}{K_+ N_+}\right), \tag{4.28}$$

where N_-, N_+ are the number densities of negative and positive ions far from the droplet. In the case of quasi-neutral atmospheric plasma $N_- = N_+$, this formula gives for the droplet charge

$$Z = \frac{rT}{e^2} \ln\left(\frac{K_-}{K_+}\right) \tag{4.29}$$

In the limit of small difference $\Delta K = |K_+ - K_-|$ for mobilities of positive K_+ and negative ions K_-, formula (4.29) gives

$$Z = \frac{rT}{e^2} \frac{\Delta K}{K}, \tag{4.30}$$

where $K = (K_+ + K_-)/2$ is the average mobility. We use these formulas in the analysis of droplet charging in a tropospheric plasma.

Let us return to Table 3.2, where a typical growth time is given for a charged microdroplet with the average radius $r = 8\,\mu\text{m}$ according to (3.1). If we base on often used mobilities of negative and positive ions in air, which are $K_- = 1.9\,\text{cm}^2/(\text{V} \cdot \text{s})$ and $K_+ = 1.4\,\text{cm}^2/(\text{V} \cdot \text{s})$ according to [76], one can obtain from formula (4.29) $Z = 40$ at the temperature $T = 275\,\text{K}$ that corresponds to the altitude $h = 2\,\text{km}$ for standard atmosphere [66]. As it follows from Table 3.2, the lifetime of a cumulus cloud with such parameters is several hours with respect to the growth of microdroplets. This more or less corresponds to reality.

Fig. 4.8 Positive (H_3O^+) and negative (OH^-) ions on the interface—the boundary between two water aerosols in different aggregate states [36]

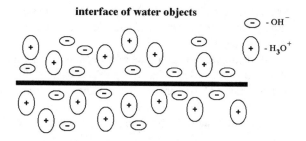

4.2.2 Charging of Water Droplets in Atmosphere

Another mechanism of the formation of charged particles in the atmosphere results from collisions of water microparticles located in different phase states. This mechanism is studied about a century ago (for example, [77]). Along with liquid and solid (ice) phases, water aerosols may contain amorphous water (hailstone) or a mixture of snow and liquid water (graupels). These types of water aerosols may coexist at altitudes of several kilometers with the temperatures between 0° and −20°. In particular, during thunderstorms which are observed in a warm season, zero temperature is attained at altitudes of several kilometers.

We give in Fig. 4.8 an interface between water particles consisting of water molecules which contain positive ions H_3O^+ and negative ions OH^-. Water is a weak electrolyte whose properties are determined by these ions. On the boundary of a water particle, a contact potential (the Workman–Reynolds potential [78]) occurs, i.e. two water particles in different aggregate states are characterized by different potentials with respect to free ions. As a result of the contact of these water particles, some ions transfer through the boundary in order to equalize the contact potentials [79, 80]. If these particles go away after their contact, they become charged. If these particles are neutral at the beginning, they obtain charges of a different sign. According to experiments [81–83], an excess of ions OH^- corresponds to droplet growth with attachment of water molecules to a water droplet, whereas an excess of ions H_3O^+ takes place in the sublimation process. Electrolyte properties of water leads to droplet charging in collisions with a surface. For example, electric discharges are observed in collisions of snowflakes with flight wires [84], and charges of cloud influence on flight electrification [85, 86].

One can expect that subsequently after collisions of water aerosols, atmospheric charges are separated as a result of faster falling of more heavy microparticles. Note that the falling velocity equals to approximately 1 cm/s for a droplet radius of $r = 10\,\mu$m and is proportional to the square of this radius ($\sim r^2$). As is seen, the separation of atmospheric charge is possible at not low sizes of falling droplets. In this case, a general scheme of electrical processes in the atmosphere [36] results from collisions of water particles in different aggregate states [78, 87, 88], mostly in collision of graupels (snow and water particles) and ice particles. Falling of charged particles leads to charge separation because of different velocities of this process.

As a result, clouds contain water particles of a certain type with larger velocities of falling. Evidently, these water particles are charged negatively, because the Earth's charge is negative.

This mechanism of charging of water and ice microparticles in their collisions was observed first in 1957 [87], where microdroplets obtained a negative charge, while microparticles of ice were charged positively. Various mechanisms of interaction may lead to this charging, but in any case the charging process results from the contact voltage at the interface between water microparticles in different aggregate states [89, 90] that lead to the charging of colliding particles [80, 81, 91–93]. Possibly, this process is responsible for microparticle charging in clouds [91–99]. In spite of different standpoints [80, 81, 91–93] for the rate of charging and the existence of empiric relations for this rate in a certain range of charging parameters [95, 100, 101], the description of this process has a qualitative character. The basic attention in this problem relates to collisions of ice particles and graupels consisting of a snow and water mixture [98, 102–104]. Therefore, we consider below another nature of the charging process due to the attachment of molecular ions of a different charge signs to water microdroplets.

We now consider another aspect of the charging problem. Let us determine the minimal charge of water microdroplets which fall to the Earth's surface under the action of the gravitation force and carry an electric charge. In considering atmospheric electric phenomena as a secondary phenomenon of water circulation through the atmosphere, we have according to formula (3.18) the flux of water microdroplets to the Earth's surface to be approximately 7×10^{16} cm^{-2}s^{-1}. The total electric current to the Earth's surface is $I = 1700$ A [67] that refers to the globe's surface $S = 5.1 \times 10^{18}$ cm^2. From this, it follows that the electron charge e relates to a number of molecules $n = 3 \times 10^{13}$. This number of molecules corresponds to a water drop of a radius $r = 6\,\mu$m. Note that in this estimation, we assume all condensed water to be contained in microdroplets of cumulus clouds. Since condensed water may be contained in other clouds with drops of another size, the above specific charge is the minimal value for this quantity, i.e. the average charge Z expressed in electron charges e for a droplet of a radius r, satisfies the criterion

$$Z > \left(\frac{r}{r_m}\right)^3 , \quad r_m = 6\,\mu\text{m} \tag{4.31}$$

If a charge Z of a droplet is created in the course of their joining, it is proportional to r^3 (r is a droplet radius), i.e.

$$Z = \left(\frac{r}{r_m}\right)^3 , \tag{4.32}$$

where r_m is the specific droplet size. One can apply this formula to water microdroplets in a cumulus cloud with parameters (3.1). According to the analysis [107], the microdroplet charge in a cumulus cloud is $Z = 27$ at the radius $r = 8\,\mu$m.

Fig. 4.9 Charge of an ice particle as a result of collisions with a riming metal rod [105, 106]

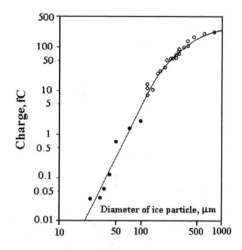

Fig. 4.10 Mobility of water ions in atmospheric air at room temperature as a function of ion molecules

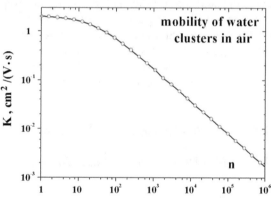

Elementary droplets with the charge $Z = 1$ have a radius $r_m = 2.7$ nm, and microdroplets of a cumulus cloud are composed from these charged microdroplets.

Figure 4.9 represents an example of the charging of an ice particle which collides with a riming metal rod of a diameter of 5 mm at the collision velocities between 2 and 9 m/s. As is seen, both the collision velocities and observed charges of ice particles exceed significantly those in cumulus clouds. Nevertheless, from the treatment of the data of Fig. 4.9, one can find $r_m = (0.48 \pm 0.03)\,\mu$m for a size radius from $r = 20\,\mu$m up to $300\,\mu$m. Note that at the largest droplet sizes of Fig. 4.10, the electric field strength on the surface of an ice particle is of the order of the breakdown electric field strength for atmospheric air.

As is seen, according to formula (4.32), the ratio of the charge Z of a growing water particle to its mass does not vary in the course of particle association. One can justify the dependence of the droplet charge Z on its size according to formula (4.32). Indeed, considering water as a weak electrolyte, one can represent the structure of a water microparticle as a system of water molecules which are grouped around a

Table 4.1 The characteristic temperature T_* at which the charge sign changes for a graupel particle which consists of a mixture of water and snow. A water graupel particle obtains a charge in collisions with ice particles. The ice density ρ_w at this transition is indicated

T_*, K	ρ_w, g/m^3	References
264.2	0.8	[109]
259.6	0.9	[110]
260.4	1.6	[111]

positive H_3O^+ ion or a negative OH^- one. Then if ΔE is the energy difference per water molecule for different aggregate states, one can estimate a typical distance between neighboring injected ions as $r_m \sim e^2/\Delta E$. From this, one can obtain the dependence (4.32) of a typical charge Z of a water particle on a radius r.

The data of Fig. 4.9 relate to the riming process, and the charging process depends on the character of droplet growth, namely, on the character of charging of a water droplet in collisions is different for growing and evaporating droplets. Moreover, the collision of two water particles located in different aggregate states may lead to a different charge sign depending on the air temperature where this process proceeds [96, 108]. In particular, Table 4.1 contains values of the air temperature at which a sign of the charge of water particles changes as a result of collision of two neutral water particles in different aggregate states.

4.3 Ions in Tropospheric Processes

4.3.1 Mobility of Small Water Ions in Atmospheric Air

Electrical processes in the atmosphere are secondary phenomena with respect to water circulation through the atmosphere. The scheme of electrical atmospheric processes is as follows [67]. On the one hand, water microparticles, mostly micro-droplets, which are located in clouds and fall down under the action of their weight, lead to the formation of charged clouds. As a result, clouds acquire an electric potential which can be transferred to the Earth through lightning or directly when charged water microparticles attain the Earth, as it takes place in winter. As a result, the Earth becomes charged. On the other hand, molecular ions which are formed under the action of cosmic rays discharge the Earth. In the course of their residence in the atmosphere, ions grow, and the current from a test ion decreases in time. These ions are lost as a result of the recombination or attachment to cloud microdroplets. The sum of the above processes determines the atmosphere's electric properties.

Thus, we separate atmospheric ions in two parts, such that the first ion type are charged water microparticles of clouds which fall down under the action of their weight and charge the Earth. For relatively small molecular ions, we refer to the other type. These ions move under the action of the Earth's electric field and discharge the Earth. In determining the boundary between these ion types, we take the average electric field strength of the Earth to be $E = 130$ V/m. In this case, the velocity of droplet falling is proportional to r^2, whereas the rate of drift of the droplet in the electric field is inversely proportional to a drop radius r. For motion downward, the following criterion has to be fulfilled [112]

$$r > 0.4\,\mu\text{m} \qquad (4.33)$$

From this, it follows that singly charged microdroplets fall down irrespective of the charge sign if their size satisfies the criterion (4.33). Ions of a less size move under the action of the Earth's electric field in such directions that their current leads to discharging of the Earth.

Thus, we take the negative charge of falling water microdroplets because the Earth is charged negatively. This falling takes place in a small part of the sky which is occupied by cumulus clouds. The basic part of the sky is free from cumulus clouds, such that this part provides discharge of the Earth as a result of currents of molecular ions under the action of the Earth's electric field. In this scheme at a certain altitude of falling microdroplets, it is possible to transfer their charge to the Earth through lightning. The average ratio of the probability of lightning and that resulted from the falling down of charged water microdroplets is 1:5 [88]. When the electric field from charged microdroplets reaches prominent objects at the Earth's surface, corona discharge (or point discharge) arises near the top of these objects.

In addition, though we assume charges of microdroplets to be negative, positive charges are possible also. According to measurements, the number of lightnings that transfer a negative charge to the ground, in 2.1 ± 0.5 times, exceeds the number of lightnings carrying a positive charge [113], and the ratio of negative current to ground to positive ones is 3.2 ± 1.2 [113]. From this, it follows that though the basic current to the Earth's surface contains negative charge, the falling of positively charged water microdroplets gives a contribution to this current.

In order to understand this problem deeper, we now consider the behavior of small ions in the atmosphere which are formed under the action of cosmic rays and grow until they recombine, as it takes place in a clear-sky atmosphere. Above, we consider the charging of microdroplets in the atmosphere and the behavior of large ions whose size exceeds remarkably the mean free path of molecules λ in the atmosphere which near the Earth's surface is $\lambda \approx 0.1\,\mu\text{m}$. We analyze below the processes involving small ions whose size is small compared with the mean free path of air molecules in the atmosphere. These ions are just responsible for the charging of water microdroplets and form an atmospheric plasma.

We first consider the mobility of small ions in an atmospheric plasma. Being guided by the mean electric field strength of the atmosphere $E = 130\,\text{V/m}$, one can determine the current density i for the Earth's discharging as

$$i = eE(K_+ + K_-)N_i, \qquad (4.34)$$

where N_i is the number density of positive and negative ions which are equal for a quasi-neutral atmosphere, K_+, K_- are the mobilities of positive and negative ions, respectively. Taking a typical number density of ions in the atmosphere

Table 4.2 Mobilities K of negative and positive ions in nitrogen [114–117] at room temperature reduced to the normal number density of molecules $N = 2.69 \times 10^{19}$ cm^{-3} and expressed in units cm^2/(V · s) [112]

Ion	K	Ion	K	Ion	K
NO_2^-	2.5	$CO_3^- \cdot H_2O$	2.1	N_2O^+	2.3
NO_3^-	2.3	N_2^+	1.9	N_4^+	2.3
CO_3^-	2.4	CO_2^+	2.2	$H^+ \cdot H_2O$	2.8
$NO_2^- \cdot H_2O$	2.4	N_2H^+	2.1	$H^+ \cdot (H_2O)_2$	2.3
$NO_3^- \cdot H_2O$	2.2	N_3^+	2.3	$H^+ \cdot (H_2O)_3$	2.1

$N_i \sim 10^3$ cm^{-3} and taking the mobilities of negative and positive atmospheric ions to be identical, one can estimate a typical mobility of ions K_i which is responsible for the discharging process

$$K_i \sim 1 \, cm^2/(V \cdot s)$$

We now consider this problem from another standpoint. Let us take simple ions which are formed at the first stage of atmospheric ionization under the action of cosmic rays. Table 4.2 contains values of the mobilities for basic negative and positive ions which are formed at the first stage of ionization of atmospheric air. These mobilities are measured at low electric field strengths. As is seen, the mobilities of considered ions differ slightly from each other, and we use below as the ion mobility the value [112]

$$K \approx 2 \frac{cm^2}{V \cdot s} \tag{4.35}$$

The comparison of data of Table 4.2 and estimation (4.35) gives that though ion growth changes the character of their transport in the atmosphere, it is not principal for estimations.

For determination of the mobility K_i of a singly charged ion consisting of a large number n of water molecules, we use formula (2.15) for the ion diffusion coefficient D_i in the Chapman–Enskog approximation and the Einstein relation [71–73] between the ion mobility and diffusion coefficient D_i

$$K_i = \frac{eD_i}{T} \tag{4.36}$$

We also use that the diffusion cross section σ_g for collision between an air molecule and a droplet of a radius r is equal to $\sigma_g = \pi r^2$, since a droplet radius r is large compared to the interaction radius a of an air molecule and a water surface, and a is of the order of an atomic value. Finally, one can obtain in the kinetic regime, where a droplet radius r is small compared to the mean free path of air molecules in air $\lambda \approx 0.1 \, \mu m$

$$K_i = \frac{3e}{8\sqrt{2\pi m T} N_a r^2} = \frac{K_o}{n^{2/3}}, \quad K_o = \frac{3e}{8\sqrt{2\pi m T} N_a r_W^2}, \quad r \ll \lambda \qquad (4.37)$$

Here, r_W is the Wigner–Seitz radius defined by formula (3.2) that for water droplets is equal to $r_W = 1.92\,\text{Å}$ [69], n is the number of water molecules of a droplet, e is the electron charge which assumes to be equal to the ion charge, m is the mass of an air molecule, T is the temperature expressed in energetic units, and N_a is the number density of air molecules. In the considering case of a water droplet located in air, we have for the mobility $K_o = 17\,\text{cm}^2/(\text{V} \cdot \text{s})$ at room temperature and normal number density of air molecules $N_a = 2.69 \times 10^{19}\,\text{cm}^{-3}$.

One can combine the mobility dependence $K_i(n)$ on the number of droplet molecules n according to formula (4.37) with the data of Table 4.2. We then obtain

$$K_i(n) = \frac{K_o}{(n_o + n)^{2/3}}, \quad K_o = 17\,\text{cm}^2/(\text{V} \cdot \text{s}), \quad n_o = 24 \qquad (4.38)$$

where at $n = 1$ formula (4.38) gives $K_i = 2\,\text{cm}^2/(\text{V} \cdot \text{s})$ in accordance with the data of Table 4.2. Figure 4.10 gives the dependence of the mobility of single-charged complex water ions in air on a number of droplet molecules.

It should be noted that on the first stage of air ionization under the action of cosmic rays, simple ions of nitrogen and oxygen are formed such as N_2^+ O_2^+, N^+, O^+, NO^+, and O_2^-, O^-. These ions grow subsequently, and complex ions are formed as a result of the attachment of molecules of type H_2O, H_2SO_4, HNO_3 to initial ions [118–125]. As a result, tropospheric air contains ions of different sizes [126–138], starting from simple ions and ending with nanometer charged droplets. However, the main contribution to the rate of ion attachment to microdroplets gives small ions with a greater mobility. Nevertheless, complex ions and ions of nanometer sizes are present in the troposphere due to a large lifetime.

4.3.2 Recombination of Ions in Atmospheric Air

The lifetime of ions in the clear-sky troposphere is determined by the recombination of positive and negative ions. We consider below this process. Let us analyze the dependence of the recombination coefficient α of small positive and negative ions on the density of air where this process proceeds. At a low pressure, this process has a pair character. For definiteness, we consider the process

$$A^+ \cdot H_2O + B^- \rightarrow (AB)^* + H_2O \qquad (4.39)$$

In this case, the released energy as a result of the recombination process is consumed on the breaking of the bond involving a water molecule.

The process (4.39) results in the transition between electron terms of the system of colliding atomic particles near the point of their intersection (for example, [63]). Let the distance between colliding atomic particles which corresponds to the intersection of these electron terms be a. The impact parameter of collision of these atomic particles ρ is connected with the distance of their strongest approach r_{min} by the relation [139]

$$\rho^2 = r_{min}^2 + \frac{e^2 \times r_{min}}{T},$$ (4.40)

in the case of the classical character of motion of atomic particles. From this, one can estimate the cross section σ_{rec} of recombination $\sigma_{rec} \sim a \times e^2/T$, and assuming the distance of the intersection of molecular terms a is of the order of an atomic value, one can estimate the recombination cross section as $\sigma_{rec} \sim 10^{-14}$ cm^2. Correspondingly, the recombination coefficient in this case is of the order of $\alpha \sim 10^{-10}$ cm^3/s.

As the number density of air molecules increases, the recombination coefficient grows. In this case, we use the Thomson theory [140] for a three body process of capture of ions and subsequent recombination in the case, where the binding energy of colliding ions is large compared to thermal energy. In thermal collisions of classical particles, one can obtain that the third particle, i.e. an air molecule, takes a part of the interaction energy between ions, and ions become bonded. One can introduce a size b of a region, where this exchange by energy is possible, on the basis of the relation

$$\frac{e^2}{b} \approx T$$ (4.41)

The parameter b is the critical radius, and the process of exchange by energy proceeds in a critical region of this size.

Under these conditions, the rate of recombination is a product of two factors, where the first one is the probability that the air molecule is located in a critical region which is of the order of $N_a b^3$. Another factor of this product is the rate of collision of an ion with air molecules with exchange by the energy of the order of T that is $v\sigma$, where v is a typical relative collision velocity, and σ is the cross section of ion–molecule collision with a remarkable exchange by energy. As a result, we obtain for the recombination rate

$$\alpha \sim b^3 v\sigma$$ (4.42)

In another limiting case with a large number density of air molecules, an interaction of recombining ions with air molecules brakes the motion of the approach of ions, and hence the recombination coefficient decreases with an increasing number density of air molecules. In this case, we use the Langevin formula (4.26) [75] for the recombination coefficient of positive and negative ions in a dense gas accounting for the motion of these ions toward each other. Then the recombination coefficient of positive and negative ions in dense air is equal if we use the Chapman–Enskog approximation for the diffusion coefficient (2.15) and mobility of ions with charges Z_+ and Z_- in air

Fig. 4.11 Dependence of the recombination coefficient for positive and negative ions in air on the number density of air molecules [141]. 1—pair ion–ion recombination; 2—three body recombination of positive and negative ions; 3—range of high pressures [141]

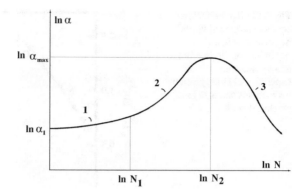

$$\alpha \approx \frac{3\sqrt{\pi}Z_+Z_-e^2}{2\sqrt{2mTN}}\left(\frac{1}{r_+^2}+\frac{1}{r_-^2}\right),\ \lambda \gg r_+, r_-, \tag{4.43}$$

where r_+, r_- are the radii of the positive and negative particle clusters, and λ is the mean free path of atoms in a gas. As it is seen, the recombination coefficient is $k_{rec} \sim 1/r^2$, where $r \sim r_+,\ r_-$. In particular, at room temperature and atmospheric pressure of air in the case $r_+ = r_- = r$, one can obtain $\alpha r^2 = 2.6 \times 10^{-20}$ cm^5/s.

Figure 4.11 represents the dependence on the number density N_a of air molecules for the recombination coefficient of positive and negative ions. We divide the range of the number density for air molecules in three parts. In the first range, at low number densities of air molecules, air molecules act weakly on the recombination process that has a pair character. In the second range, the recombination of ions has the three body character. Then the recombination coefficient of positive and negative ions is proportional to the number density of air molecules and is estimated by formula (4.42).

One can estimate the transiting number densities of air molecules assuming for simplicity that $a = a_o$ (a_o is the Bohr radius) and taking the cross section of polarization capture for collisions between an ion and air molecule. Then we obtain

$$N_1 \sim a_o(T/e^2)^{5/2}\beta^{-1/2},$$

where β is the polarizability of an air molecule, and we assume masses of ions and air molecules to have the same order of magnitude.

The maximal recombination coefficient is expected at the number density

$$N_2 \sim \frac{T^{3/2}}{e^3\sqrt{\beta}},$$

and the maximum recombination coefficient is estimated as

Fig. 4.12 Recombination
coefficient of positive and
negative ions in air
depending on air pressure in
accordance with
experimental [142] given by
open circles and data [143]
by closed circles

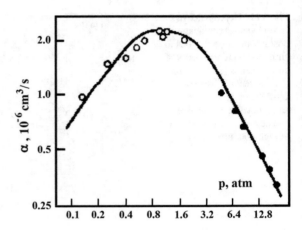

$$\alpha_{max} \sim v b^2 \sim e^4 \mu^{-1/2} T^{-3/2} \tag{4.44}$$

On the basis of these estimations, we have for typical values of parameters under
consideration [141] : $N_1 \sim 10^{17}$ cm^{-3}, $N_2 \sim 10^{20}$ cm^{-3}, and $\alpha_{max} \sim 10^{-6}$ cm^3/s.
Figure 4.12 gives measured rate constants [142, 143] for the recombination of positive
and negative ions in air as a function of the air pressure. The maximum value of the
recombination coefficient corresponds to pressures of the order of the atmospheric
one.

We now concentrate on large complex ions in atmospheric air whose size is small
compared to the mean free path of molecules in air $\lambda \sim 0.1\,\mu$m, i.e. the number of
water molecules in this complex ion $n \ll 10^8$. At such sizes, the ion charge is equal
to the electron charge $\pm e$. Next, the number density of ions N_i is relatively small,
i.e. a distance from the test ion to the nearest one $\sim N_i^{-1/3}$ is large compared to the
mean free path λ_i of ions in air. On the basis of formula (4.43), one can estimate the
mean free path of complex ions in atmospheric air as

$$\lambda_i \sim \frac{10^{-4}\,\text{cm}}{n^{2/3}} \tag{4.45}$$

Hence, the criterion $N_i \lambda_i^3 \ll 1$ is fulfilled, at last, for $N_i \ll 10^{12}$ cm^{-3}, and the recombination process proceeds in the diffusion regime, where the recombination coeffi-
cient is determined by the Langevin formula (4.26)

$$\alpha = 4\pi e (K_- + K_+), \tag{4.46}$$

where K_-, K_+ are the mobilities of negative and positive ions correspondingly.

4.3.3 Ions in Clear-Sky Atmosphere

In considering the processes of atmospheric electricity, we separate the atmosphere into two parts. There are rare clouds in the main part of the atmosphere with a clear sky. In this atmosphere part, water microdroplets are not important in electric processes, and processes of Earth's discharging proceed with the participation of ions formed under the action of cosmic rays. In another part of the atmosphere with cumulus clouds, charging of the Earth takes place as a result of the falling of charged water microdroplets. This atmosphere part covers a small part of the Earth's surface, but contains the most part of condensed atmospheric water. We consider below the processes in each part of the atmosphere separately.

In this consideration, we take into account that in a clear sky of the atmosphere has a low density of water microdrops. Hence this part of the atmosphere is responsible for the Earth's discharging as a result of ion currents in the atmospheric electric field. Hence, in this atmosphere part ions are formed as a result of the atmospheric ionization under the action of cosmic rays and recombine in collision processes of positive and negative ions. Until ions are located in the atmosphere, these ions grow by the attachment of water molecules to ions. Below, we take into account these processes in a clear sky.

The key place in the analysis of the ion kinetics in this part of the atmosphere is the balance equation for the ion number density N_i that has the form

$$\frac{dN_i}{dt} = M - \alpha N_i^2, \tag{4.47}$$

where M is the ionization rate of the atmosphere, $\alpha \approx 2 \times 10^{-6} \, \text{cm}^3/\text{s}$ is the recombination coefficient in accordance with Fig. 4.11. The character of atmospheric ionization by cosmic rays is represented in Fig. 4.5. As is seen, the maximum ionization rate of atmospheric air is observed at altitudes of 11–15 5 km. In this altitude range, which is the Pfotzer maximum [40]), the air ionization rate is approximately $30 \, \text{cm}^{-3}\text{s}^{-1}$ [38, 39, 41, 42]. The total air ionization rate per unit area of the Earth's surface is $4.5 \times 10^7 \, \text{cm}^{-2}\text{s}^{-1}$ [43]. The ratio of these values gives a typical range of atmospheric altitudes $h \sim 10 \, \text{km}$, that is the size of the region where air ionization proceeds.

Near the Earth's surface, the ionization rate of air is determined both by cosmic rays and as a result of Earth's radioactivity, and this value attains $10 \, \text{cm}^{-3}\text{s}^{-1}$ [45]. At altitudes of $200 - 300 \, \text{m}$, this rate is approximately $\sim 4 \, \text{cm}^{-3}\text{s}^{-1}$ [47] and decreases with an altitude increase. Because the main contribution to atmospheric electricity follows from altitudes of several km, we take the rate of atmospheric ionization by cosmic rays as $M \sim 5 \, \text{cm}^{-3}\text{s}^{-1}$ (see also Fig. 4.5). Taking according to Fig. 5.3 $\alpha \sim 10^{-6} \, \text{cm}^3/\text{s}$, one can obtain a typical number density of ions $N_i = 2 \times 10^3 \, \text{cm}^{-3}$, and a typical recombination time $\tau_{\text{rec}} \sim (M\alpha)^{-1/2} \sim 500 \, \text{s}$, if an ion size does not change in the course of recombination.

In addition, the maximum number density of ions $N_i \approx 6 \times 10^3 \, \text{cm}^{-3}$ is observed at altitudes of 11–15 km with maximal ionization, as it follows from the balance

(4.47). A typical lifetime is equal from this equation $\tau \sim N_i/\alpha \sim 3$ min at these altitudes. During this time, an ion travels a path of about hundred meters. It should be noted that the variability of the intensity of galactic and solar cosmic rays causes a change of these estimations for ionization of the atmosphere [29].

It should be noted that in the fine part of the atmosphere, water vapor is non-saturated for its most part. Therefore, in spite of a variety of complex ions in the atmosphere, water microdroplets are not formed there. Hence, we assume ions in this atmosphere part to have the mobility $K_i \approx 2\,cm^2/(V \cdot s)$ at the first stage of their evolution. We now make estimations on the basis of measured discharge current in a fine atmosphere. The mean current density in the fine atmosphere is $2.4\,pA/m^2$ over the land and $3.7\,pA/m^2$ over oceans according to observed data [144], which corresponds to the average flux of ions $i = 3 \times 10^3\,e/(cm^2 \cdot s)$. From this, we have for the specific density of the atmospheric discharging current

$$i = 2eK_i N_i E, \tag{4.48}$$

and the factor 2 accounts for the current of positive and negative ions in two directions. From this, one can determine the number density of ions N_i which create this current

$$N_i = \frac{2i}{K_i E} \sim 10^3\,cm^{-3} \tag{4.49}$$

Here, $K_i = 2\,cm^2/(V \cdot cm)$ is the ion mobility, and $E = 1.3\,V/cm$ is the electric field strength. As is seen, this estimation accords with that followed from (4.47) of the ion balance.

4.4 Electric Processes in Cumulus Clouds

4.4.1 Ions in Cumulus Clouds

In consideration of the atmospheric electric properties, we separate the atmosphere into two parts, so that in the most part of this, in a fair atmosphere, discharging of the Earth occurs due to ion currents under the action of the Earth's electric field. Charging of the Earth proceeds owing to cumulus clouds are located which contain the basic part of condensed atmospheric water and cover a small part of the Earth's surface. The only mechanism of charge separation in the atmosphere results from gravitation falling of charged droplets toward the Earth's surface.

Note that the velocity of gravitation falling of particles in the atmosphere is proportional to the square of a particle radius. Cumulus clouds are characterized by the maximal water density and typical parameters of water microdroplets of cumulus clouds are given by formula (3.1). We neglect processes involving water microdroplets for rare clouds, so that only cumulus clouds partake in the charging of the Earth. For simplicity, we assume that all microdroplets of a cumulus cloud have an

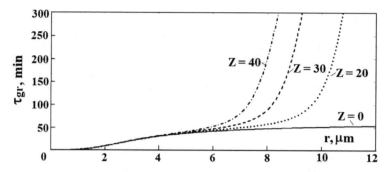

Fig. 4.13 Total time of growth of water droplets up to an indicated size in a cumulus cloud if an average number density of bound atoms in a space is equal to that of a cumulus cloud according to (3.1). This gives $N_b = 7.2 \times 10^{16}\,\text{cm}^{-3}$ at atmospheric pressure and temperature $T = 273\,\text{K}$. An indicated droplet charge Z is reached at a droplet radius $r = 8\,\mu\text{m}$

unidentical radius (3.1) and charge Z. The droplet charge prevents microdroplets from their joining, and one of methods to find a typical droplet charge Z of a microdroplet is based on the lifetime of cumulus clouds [107]. Under the above conditions, $Z = 27 \pm 2$ [107] if the lifetime of a cumulus cloud is between 3 and 12 h.

A more precise method to estimate the microdroplet charge on the basis of the lifetime of a cumulus cloud consists in the kinetic analysis of the growth of microdroplets. The doubling time at a given radius of a microdroplet and for its indicated charge is represented in Fig.3.13, and in addition to this we give in Fig. 4.13 the total growth time $\tau_{gr}(r)$ up to an indicated droplet radius for some charges of a microdroplet. These times characterize the growth kinetics for charged microdroplets accounting for coalescence and gravitation growth mechanisms. As is seen, at some sizes, where electrostatic repulsion of microdroplets becomes essential, the growth process of microdroplets slows down, so that the growth process at a certain droplet size is stopped practically.

In addition to this, we give in Table 4.3 some parameters of growth of water microdroplets if the charge density Z/r^3 (Z is the microdroplet charge, r is the microdroplet radius) is conserved in the course of the growth process. This growth time $\tau_{gr}(r)$ of a water microdroplet up to a radius r is defined according to the following formula which takes into account the joining of microdroplets as a result of coalescence and gravitation joining in accordance with formula (3.48)

$$\tau_{gr}(r) = \int\limits_0^r \frac{3R^2 dR}{r_w^3/\tau_c + R^4 \exp\left(-\frac{Z(R)^2 e^2}{2RT}\right)/(\tau_{gr} \cdot r_w)} \tag{4.50}$$

where $Z(R) \sim R^3$ is the charge of a droplet of a radius R.

In this consideration, we assume that a cumulus cloud, inside which the growth process occurs, to be uniform and parameters of microdroplets are given by (3.1) at a moment when the droplet has a radius $r = 8\,\mu\text{m}$. During the growth time, a

Table 4.3 A time τ_{gr} of growth of water microdroplets up to a radius of $r = 8\,\mu m$ at the temperature $T = 273\,K$ and atmospheric pressure according to formula (4.50); Δh is the altitude which a microdroplet passes through this time. The ratio q/m of the droplet charge to its mass is conserved in the course of droplet growth, and Z is the microdroplet charge at its radius $r = 8\,\mu m$. The indicated relative difference $\Delta D_i / D_i$ between the diffusion coefficients of negative and positive ions determines the charge Z of water microdroplets in accordance with formula (4.29)

Z, e	τ_{gr}, h	q/m, 10^{-10} C/g	$\Delta D_i / D_i$	Δh, m
0	0.80	0	0	10
2	0.80	1.5	0.015	10
5	0.81	3.7	0.038	10
10	0.83	7.5	0.076	11
15	0.88	11	0.11	12
20	0.96	15	0.15	15
25	1.1	19	0.19	22
30	1.5	22	0.23	35
40	3.3	30	0.30	110

microdroplet falls down with the velocity $w(r) \sim r^2$ (r is the droplet radius) and $w = 0.8\,cm/s$ at $r = 8\,\mu m$. In the course of falling, a microdroplet passes a path $\Delta h(r)$

$$\Delta h(r) = \int_0^r \frac{3R^2 dR \cdot w(R)}{r_W^3 / \tau_c + R^4 \exp\left(-\frac{Z(R)^2 e^2}{2RT}\right) / (\tau_{gr} \cdot r_W)} \tag{4.51}$$

Assuming that all the microdroplets of a cumulus cloud have an identical radius $r = 8\,\mu mn$, we define the charge Z of Table 4.3 as that is attained by a microdroplet at an indicated radius. Note that the ratio of the total current toward the Earth $I = 1700\,A$ to the total mass of evaporated water per unit time 1.5×10^{13} g/s from the Earth's surface is equal to $q/m = 1.2 \times 10^{-10}$ C/s. As it follows from Table 4.3 data, this value reaches at low specific charges which do not influence the growth process.

We assume above that the charge of a microdroplet results from the attachment of ions to it, and its value is given by formula (4.29). In particular, according to measurements [76], the mobility of positive ions at the temperature $T = 288\,K$ and atmospheric pressure is $K_+ = 1.4\,cm^2/(V \cdot s)$, whereas that for negative ions it equals $K_- = 1.9\,cm^2/(V \cdot s)$. Under these conditions, formula (4.29) and equilibrium conditions give the negative charge of a microdroplet of a radius $r = 8\,\mu m$ as $Z = 42$. As it follows from the lifetime of a cumulus cloud, a typical charge of a microdroplet of a cumulus cloud (3.1) is $Z = 27 \pm 2$ [107].

In considering atmospheric electricity as the secondary phenomenon of water circulation through the atmosphere, one can connect the electric current density of the Earth's charging with the flux of evaporated molecules that is equal to the flux of precipitated water. Taking the total rate of water evaporation from the Earth's surface to be $dm/dt = 1.5 \times 10^{13}$ g/s and the electric current to the Earth's surface to be $I = dq/dt = 1700$ A, where q is a current charge, one can obtain for the specific charge

$$\frac{dq}{dm} = 1.1 \times 10^{-10} \, \text{C/g} \tag{4.52}$$

In the case of the average microdrop charge $Z = 27$, one can obtain the specific rate of the Earth's charging as $dq/dm = 2 \times 10^{-9}$ C/g. Taking into account that the Earth's charging is determined by water microdroplets, i.e. by the water condensed phase, one can estimate the part of the condensed phase in the atmosphere as ~6%.

Note that the microdroplet charge depends on ion sorts which are formed in atmospheric air of a cumulus cloud under the action of cosmic rays, and these sorts, in turn, are determined by air additions on the basis of NH_3, H_2SO_4, HNO_3 which are present in this air. Therefore, the microdroplet charge may be different, and even it can be positive. For this reason, though microdroplets are negatively charged usually, more seldom they may be charged positively. In particular, according to measurements [113], the ratio of a number of lightnings which transfer a negative charge to the Earth's surface to that with the negative charge is 2.1 ± 0.5, whereas the ratio of lightning currents with negative and positive charges is 3.2 ± 1.2 [113]. Because lightnings transfer the cloud charge to the Earth's surface, the basic charge of clouds is negative. Hence, lightnings transfer a negative charge to the Earth's surface mostly and provide the Earth's charging [67, 145] negatively.

We now consider another aspect of growth of charged microdroplets in a cumulus cloud. Taking the balance equation for the number density N_i of ions (4.47), we add the term with ion attachment to water microdroplets in this balance equation, which takes the form

$$\frac{dN_i}{dt} = M - \alpha N_i^2 - \nu_{at} N_i, \tag{4.53}$$

where ν_{at} is the attachment rate of ions to water microdroplets. The expression for the attachment rate on the basis of the Smoluchowski formula (2.30) has the form

$$\nu_{at} = 4\pi D_i r N_d \tag{4.54}$$

Here, r is a microdroplet radius, D_i is the diffusion coefficient of ions in atmospheric air, and N_d is the number density of microdroplets. We take parameters of water microdroplets in a cumulus cloud according to (3.1), i.e. $r = 8 \, \mu\text{m}$, $N_d = 10^3 \, \text{cm}^{-3}$, and the average mobility of ions under these conditions $K_i = 2 \, \text{cm}^2/(\text{V} \cdot \text{s})$. The latter gives according to the Einstein relation $D_i = 0.05 \, \text{cm}^2/\text{s}$, and the attachment rate is $\nu_{at} = 0.5 \, \text{s}^{-1}$.

From the balance (4.53) it follows that under real conditions in a cumulus cloud, the loss of ions results from the attachment of ions to water microdroplets. Taking as above $M = 10 \, \text{cm}^{-3}\text{s}^{-1}$, one can obtain from equation $M = \nu_{at} N_i$ for the equilibrium number density of ions $N_i = 20\,\mathrm{cm}^{-3}$. A typical time of equilibrium establishment under these conditions, as it follows from (4.53), is $\tau_i \sim 2\mathrm{s}$.

We note one more peculiarity of the charging process under consideration. Taking for definiteness the droplet charge $Z = 27$ [107], one can obtain the equilibrium number density of an excess negative charge $N_Z \sim 3 \times 10^4\,\mathrm{cm}^{-3}$, and this negative charge is bonded with water microdroplets. But the average equilibrium number density in the region where these water droplets are located is $N_i \approx 20\,\mathrm{cm}^{-3}$, and the difference of the number densities of positive and negative ions is equal to $\Delta N_i = N_+ - N_- \approx 4\,\mathrm{cm}^{-3}$. As is seen, fluxes of positive and negative ions to the surface of water microdroplets are almost identical. It is important that the equilibrium charge of water microdroplets exceeds approximately $\sim 10^3$ times that from a surrounding space. Hence, it is necessary to replace $\sim 10^3$ times air which surrounds the region occupied by water microdroplets in order to provide their equilibrium charge.

The above character of processes in the region occupied by water microdroplets takes place if the change by air between this region and surrounding ones proceeds slowly. Evidently, the criterion of a slow character of air replacing is given by

$$\tau_{\mathrm{exch}}\nu_{at} \gg 1, \tag{4.55}$$

where τ_{exch} is the exchange time by air between the region occupied by microparticles and surrounding ones. Inserting in formula (4.54) a typical microdroplet radius $r = 8\,\mu\mathrm{m}$ and their number density in a cumulus cloud $N_d = 10^3\,\mathrm{cm}^{-3}$ according to formula (3.1), and the diffusion coefficient of ions $D_i = 0.05\,\mathrm{cm}^2/\mathrm{s}$, we have for a typical rate of ion attachment to microdroplets $\nu_{at} \approx 0.5\,\mathrm{s}^{-1}$. An average velocity of horizontal atmospheric winds is $v_w = 5\,\mathrm{m/s}$ [146–148]. Hence, a slow regime of droplet charging in a cumulus cloud takes place if the size of the region occupied by water microdroplets inside a cumulus cloud L satisfies the criterion $L \gg v_w/\nu_{at} \sim 10\,\mathrm{m}$.

We thus obtain the blanket charge structure of a cumulus cloud, where regions occupied by charged water microdroplets alternate with those without microdroplets. At the first stage of cloud ripening, regions occupied by microdroplets are charged negatively, whereas surrounding regions without microdroplets are charged positively. In the case under consideration [107] with the average charge $Z = 27$ in electron charge units for an individual microdroplet, the charge number density of a region occupied by water microdroplets is

$$N_Z = 3 \times 10^4\,\mathrm{cm}^{-3} \tag{4.56}$$

From this, one can estimate the number density of positive ions in surrounding regions N_+ as

$$N_+ \sim N_Z \cdot \frac{v_w/\nu_{at}}{L}, \tag{4.57}$$

where v_w is a typical velocity of winds which separate positively and negatively charged regions, and L is a typical size of a region with negatively charged water microdroplets. Subsequently, separation of charges takes place both as a result of the falling of charged microdroplets and under the action of winds, both horizontal and vertical ones. After this a cumulus cloud or its lower part becomes negatively charged.

Let us consider in detail the charging of water microdroplets through attachment of molecular ions which are formed under the action of cosmic rays, if subsequently these ions are removed from the region occupied by water microdroplets. As a result, the region with negatively charged water microdroplets is surrounded by a more large region which contains an excess of positive ions. Indeed, due to the difference of mobilities for molecular ions located in the atmosphere, it is possible to create a certain negative charge on each microdroplet, and we take it as $Z = 27$ in accordance with properties of a cumulus clouds [107]. But the number density of molecular ions in the atmosphere is low, and in order to provide this charge for each microdroplet, it is necessary to renew air in the region of microdroplets many times. This may proceed as a result of two alternatives, namely, under the action of winds and as a result of falling of water microdroplets in the atmosphere. We analyze below these processes.

We first consider the departure of air inside the region occupied water microdroplets outside this region under the action of wind gust. Let the air velocity to vary at the beginning from 0 up to v_o. Air inside the microdroplet region is captured by this wind gush prompt, but microdroplets will move with a new velocity through some time. As a result of these processes, air displaces with respect to microdroplets at a distance L that is given by

$$L = \int [v_o - v(t)]dt,$$

where $v(t)$ is a current microdroplet velocity. This velocity is given by motion equation on the basis of Stokes formula

$$\frac{M dv}{dt} = 6\pi \eta r v, \tag{4.58}$$

where M is a microdroplet mass, r is a microdroplet radius, v is the relative velocity between the microdroplet and air, and $\eta = 1.7 \times 10^{-4} \text{g}/(\text{cm} \cdot \text{s})$ is the air viscosity at an altitude ~ 3 km of standard atmosphere. The solution of (4.55) has the form

$$v = v_o \left[1 - \exp(-t/t_o)\right], \quad t_o = \frac{2r^2 \rho}{3\eta} \tag{4.59}$$

From this, it follows for a shift L of air with respect to microparticles $L = v_o t_o$. Taking the microdroplet radius $r = 8\,\mu\text{m}$ on the basis of (3.1), one can obtain under these conditions $t_o = 2$ ms. Taking a typical wind velocity $v_o \sim 5$ m/s, one can obtain a typical air shift in this process $L \sim 1$ cm. This mechanism leads to the blanket structure of a cumulus cloud consisting of water microdroplets.

Another mechanism is realized in the course of the falling of charged microdroplets and their charging as a result of the attachment of negative and positive molecular ions to them. In this manner, the charging of water microdroplets transfers the atmospheric charge. The falling velocity for the average radius $r = 8\,\mu$m of a water microdroplet (3.1) is equal to $w_g \approx 1$ cm/s according to formula (2.17). A typical separation time for a characteristic cloud size $L \sim 100$ m is measured in hours.

In conclusion of this analysis, it should be noted that the above estimations were based on typical parameters of a cumulus cloud and some conditions there. In reality, a cumulus cloud of a certain size may carry a positive charge. Hence, the above yield parameters of cumulus clouds are estimations. Nevertheless, from this analysis one can conclude that at the first stage of charging, a cumulus cloud may have a blanket structure with alternative negatively and positively charged regions, or to be consisting of a negative layer with water microdroplets below and a positively charged layer above it, where an excess of positive molecular ions is observed. Subsequently under the action of horizontal or vertical winds, these regions may be removed from each other, and a cumulus cloud as a whole becomes negatively charged, whereas the positive charge is spread over a space. The mass of condensed water due to which the separation of charges proceeds in the atmosphere is several percents of the total atmospheric mass.

4.4.2   Processes in Cumulus Clouds Involving Water Microdroplets

Since key electric processes in the atmosphere proceed inside cumulus clouds, we consider below some processes in cumulus clouds which allow one to represent a physical picture of electric processes in the troposphere. The creation of the electric voltage of the atmosphere results from the falling of water microdroplets and proceeds in cumulus clouds which contain the main amount of atmospheric condensed water. We consider electric processes in the atmosphere as a secondary phenomenon of water circulation. The connection of these processes is characterized by the parameter (4.52) that is the ratio of the electric charge q transferred by water to its mass M, and this ratio is equal to

$$\frac{q}{M} = 1.1 \times 10^{-10} \text{C/g} = 2 \times 10^{-14} \text{e/molecule}, \qquad (4.60)$$

that is one electron charge relates to 5×10^{13} molecules.

From this, one can estimate the efficiency ξ for the process of formation of the Earth's electric field as a result of the falling of charged water microdroplets. In this process, charged microdroplets fall down under the action of their weight in the

Fig. 4.14 Typical
distribution of charges in
thunderstorm cloud [149,
150] that is constructed on
the basis of measurements of
cloud electric fields in South
Africa. In this case, the
positive charges are
$P = 40\,C$, $p = 10\,C$; the
negative charge is
$N = -40\,C$

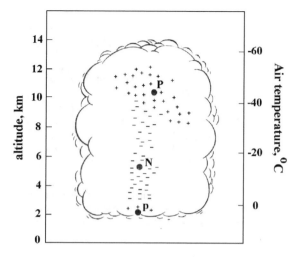

opposite direction with respect to the electric field, and the efficiency of this process
for charge transport is determined as

$$\xi = \frac{ZeE\delta h}{mg\delta h} = 3 \times 10^{-5}, \tag{4.61}$$

where $ZeE\delta h$ is the electric energy which is accumulated as a result of the passing
of a distance δh for a charged particle of a charge Ze against an electric field of
the strength E, and $mg\delta h$ is the gravitation energy after descent on an altitude δh
for a microdroplet of a mass m. We take above the charge $Z = 27$ for a typical
microdroplet (3.1) of a cumulus cloud, and g is the free fall acceleration. As is seen,
a small portion of the energy during water circulation through the atmosphere is
consumed on the creation of the Earth's electric field.

We also show that a cumulus cloud takes the blanket structure in the course of the
charging of microdroplets, and at this stage of evolution of a cumulus cloud, regions
with negatively charged water microdroplets alternate with regions without water
microdroplets, where the positive charge of molecular ions dominate. An example of
the charge distribution in a cumulus cloud is represented in Fig. 4.14. This confirms
the blanket character of charge distribution in a cumulus cloud, so that positive and
negative charges are separated inside it both in horizontal direction and in a vertical
one (Fig. 4.15).

In the above consideration, we combine processes of growth of water microdroplets in a cumulus cloud and their charging, but these processes were separated
from the process of creation of the blanket structure of a cumulus cloud. But in reality,
these processes proceed simultaneously, and the formation of cumulus clouds takes
place in restricted regions with ascending streams of wet warm air. Therefore, in contrast to the above consideration, the amount of condensed water increases in time.
Figure 4.15 represents a scheme of water penetration in a cumulus cloud, and this

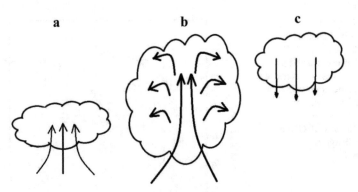

Fig. 4.15 Character of evolution of the ascending stream of warm wet air: **a** The first stage of water evaporation and condensation. **b** Simultaneous growth and charging of water microdroplets. **c** Transition of charges from microdroplets in a space that leads to rain and lightning [151, 152]

scheme is constructed on the basis of measurements [151, 152]. Thus, on the basis of observations of atmospheric electricity and measurements of some parameters, one can combine a general physical picture of corresponding processes. Below, we give this picture as it was, at least, 50 years ago (for example, [150, 153–160]) and our goal is to represent some estimations for these processes from the microscopic standpoint on the basis of this description.

After formation of charged cumulus clouds, we have two scenarios of their subsequent development. In the first case (approximately 80% of cases), charged water microdroplets descend and transfer their charge to the Earth, while in the second case lightning occurs, and the lightning current transfers partially a cloud charge to the ground. As a rule, not only the Earth is charged negatively in these processes (approximately 80% of the current to the Earth), but also the transfer of positive charge gives a certain contribution to the current to the Earth (approximately 20%). Next, if charged water microdrops of a cumulus cloud fall down, this may cause corona discharge near outstanding objects. In particular, sailors call this phenomenon, which consists in the glowing of the mast and other outstanding objects of the ship, as Sant Elmo fires. On land, corona discharge is observed on branches of trees. In mountains, metallic objects, as hooks, ice axes et.al., become ringing in a thunderstorm weather under the action of atmospheric currents due to charged water microdroplets.

Another way of discharging of cumulus clouds proceeds due to lightning. As it follows from observations, formation of lightning requires special conditions. In any case, the surface temperature cannot be low, and lightnings are observed mostly in summer. This allows one to consider lightning as an electric instability in the process of descent of charged microdroplets of a cumulus cloud to the ground in an atmosphere that is not cold. Another property of this character of atmospheric processes is that a cumulus cloud is discharged, and therefore the process of growth of water microdroplets is accelerated. As a result, water microdroplets are converted into rain drops. Hence, lightning is associated with rain.

We first estimate a minimal size of a bunch of charged water microdroplets which give the electric field strength of the thunderstorm atmosphere that is approximately $E_o = 200 \text{V/cm}$. In this estimation, we are guided by above typical parameters, that is the microdroplet charge $Z = 27$, a typical radius $r = 8 \mu\text{m}$, and the number density of microdroplets $N_d = 10^3 \text{cm}^{-3}$. This corresponds to the charge number density (4.56) for an element of a cumulus cloud occupied by charged microdroplets. One can obtain for the electric field strength E near the surface of a ball of a radius R that follows from equation

$$E = \frac{4\pi R}{3} N_Z e \qquad (4.62)$$

In particular, the electric field strength $E_o = 200 \text{ V/cm}$ is attained at the surface of such a ball at its radius $R = 100 \text{ m}$. In addition, we determine the electric field strength which is created by a long flat cloud of thickness l. In this case, the electric field strength is equal to

$$E = 4\pi\sigma = 4\pi e N_Z l, \qquad (4.63)$$

where σ is the charge surface density, and at the above parameters we have for the cloud charge per unit area of the atmospheric column and for the specific electric field strength

$$\sigma = \frac{E_o}{4\pi} = 1 \times 10^8 \frac{e}{\text{cm}^2} = 2 \times 10^{-11} \frac{C}{\text{cm}^2}, \frac{E}{l} = 0.05 \frac{V}{\text{cm}^2} \qquad (4.64)$$

From this, one can estimate the thickness of a continuous cumulus cloud as $l = 40 \text{ m}$. Taking into account that cumulus clouds have a blanket structure, one can find the thickness of cumulus clouds in the thunderstorm weather to be of hundreds meters.

One can confirm one more parameter of cumulus clouds during the thunder storm. The electric potential of these clouds with respect to the Earth's surface is 20–100 MV [161]. This electric potential is realized at the average electric field strength $E_o = 200 \text{V/cm}$ if clouds are located at the altitude $(1–5)$ km. Because the length of cumulus clouds exceeds its altitude over the Earth's surface, from this it follows that typical lengths of cumulus clouds during thunderstorm are of the order of 10 km. Next, during one lightning flash the charge $(4–5)$ C transfers from clouds to the ground [162–166], and 3–4 subsequent flashes are realized in one breakdown on average. From this it follows the cloud area to be $S \sim 100 \text{ km}^2$, i.e. a length of clouds which partake in the creation of lightning breakdown is of the order of 10 km. We also note that both the lightning current and repetition of electric breakdown through the same channel are realized due to charge transport on the Earth because the Earth's conductivity is higher than that of the atmosphere.

4.4.3 Stability of Cumulus Cloud

Cumulus clouds are of importance for atmospheric electricity because they are responsible for the charging of the Earth as a result of gravitation falling of charged water microdroplets. As a matter of fact, a cumulus cloud is a suspension with water microdroplets located in atmospheric air. This system contains also water vapor that consists of water molecules which are found in equilibrium with water microdroplets. Therefore, the number density of water molecules in cumulus clouds is equal to that of the saturated one, i.e. to the number density of molecules at the saturated vapor pressure for the air temperature.

This system is unstable with respect to the process of association of microdroplets in larger droplets, and then water in the form of water droplets is removed from air as a result of gravitation fall of rain droplets. The above analysis of the growth of water microdroplets in atmospheric air shows (see Figs. 3.13 and 4.13) that the lifetime of charged microdroplets with respect to their growth may be large enough because droplet charges prevent them from a contact. Therefore, the growth lifetime of charged microdroplets exceeds significantly that of neutral microdroplets. Hence, below we ignore this channel of decay of cumulus clouds.

In the course of evolution of cumulus clouds, the equilibrium is supported between water microdroplets and free water molecules, so that the number density of free water molecules is equal to that of the saturated one. A change of the air temperature or the number density of water molecules returns this system to the equilibrium by means of attachment of excess molecules to microdroplets. Let us estimate this time being guided by average parameters (3.1) of microdroplets and an altitude of $h = 3$ km. For the model of standard atmosphere, this corresponds to the air temperature $T = 268$ K, and the saturated number density of water molecules at this temperature according to Table 3.1 equals $N_w = 1.1 \times 10^{17}$ cm^{-3}, while the number density of bound water molecules in microdroplets is equal to $N_b = 7 \times 10^{16}$ cm^{-3}.

A typical time τ_{at} of equilibrium establishment between free and bound water molecules in atmospheric water is given by the Smoluchowski formula (2.30) and is equal to

$$\frac{1}{\tau_{at}} = 4\pi D_w N_d r, \qquad (4.65)$$

where $D_w = 0.22$ cm^2/s [167, 168] is the diffusion coefficient for water molecules in air at this altitude, and $N_d = 10^3$ cm^{-3} is the number density of microdroplets in an average cumulus cloud (3.1). From this, we have $\tau_{at} \sim 0.5$ s. Such times characterize the attachment of excess free molecules to microdroplets or evaporation of free water molecules from the microdroplet surface. Thus, cumulus clouds react fast compared to their lifetime to small changes of their parameters .

It is clear that a cumulus cloud disappears at its heating which leads to the evaporation of microdroplets. Under the considered conditions, this is realized if the system is heated up to the temperature where the total number density of water molecules

$N_t = N_w + N_d n$ is equal to the saturated number density at a new temperature. According to data of Table 3.1, this can result from the temperature change of this cumulus cloud by approximately 7 K. But this cannot be reached by this temperature change because energy is required to transform water microdroplets into free water molecules. Evidently, this requires an additional heating by the temperature

$$\Delta T = \frac{\varepsilon_b * N_d * n}{c_p N_a} \approx 5K \tag{4.66}$$

Here, $\varepsilon_b \approx 0.43\,eV$ is the binding energy for a water molecule in a liquid microdroplet, $n = 7 \times 10^{13}$ is the number of water molecules in one microdroplet, and $c_p = 7/2$ is the heat capacity per one air molecule. From this, we obtain that under the above conditions microdroplets of a cumulus cloud evaporate if it is heated by temperature of approximately 12 K.

4.4.4 Thermal Charge Release for Microparticles

In considering the charging process of a water microdroplet as a result of attachment of negative and positive atmospheric ions to its surface, it was assumed a small equilibrium charge of a microdroplet (4.29) compared to a possible one. In other words, let us remove positive ions, and then the equilibrium is established between molecules B, negative ions B^-, and a charged microdroplet according to equation

$$A_n^{-Z-1} + B \leftrightarrow A_n^{-Z} + B^-, \tag{4.67}$$

Here, A_n^{-Z} means a microdroplet which consists of n water molecules and its negative charge is equal to Z electron charges. Evidently, this equilibrium charge exceeds significantly that (4.29) resulted from the attachment of atmospheric ions to this microparticle.

Note the character of development of atmospheric electric breakdown when charged water microdroplet descend to low warm layers of the atmosphere. Usually, electric breakdown of the atmosphere is accompanied by rain that reaches the ground through several minutes after the first flash [162–166]. This means that simultaneously with the atmospheric breakdown, water microdroplets lose their charge that leads to their fast growth up to the size of rain drops. One can conclude from this that a temperature increase causes a shift of the charge equilibrium (4.67) that leads to a partial charge release from water microdroplets. We analyze below the conditions under which charged microdroplets lose the charge as a result of the temperature increase.

Using the analogy with charged clusters, we use the analysis of charge equilibrium for clusters [69, 112]. Then the ratio of probabilities P_{-Z} and P_{-Z-1} that a microdroplet contains a charge $-Z$ and $-Z - 1$ correspondingly is given by the Saha distribution that in this case has the form [69, 112]

$$F(Z) = \frac{P_{-Z}N_-}{P_{-Z-1}} = G_Z^{-1} \cdot \xi \cdot \exp\left(-\frac{\varepsilon_Z}{T}\right), \ \xi = \left(\frac{mT}{2\pi\hbar^2}\right)^{3/2} \quad (4.68)$$

Here, N_- is the number density of free negative ions, G_Z is the statistical weight for negative ions bonded with the surface of this microdroplet, m is the mass of the negative ion, ε_Z is the binding energy for a negative ion bonded with the microdroplet surface if its charge is Z, and T is the air temperature. Here, ions are located in thermodynamic equilibrium with atmospheric air, and we estimate below the typical parameters of this equilibrium.

In considering a water microdroplet as a dielectric system, one can obtain that negative ions may be found at certain knots or active centers of its surface. A number of such knots for location of negative ions is large compared to a number of bonded negative ions at the microdroplet surface. Under such conditions, the microdroplet statistical weight G_Z is estimated as

$$G_Z \sim \left(\frac{r}{r_d}\right)^2, \quad (4.69)$$

where r is a microdrop radius, and r_d is a typical distance between neighboring bonding knots of the microdroplet. Assuming this distance to be of the order of the Wigner–Seitz radius for water molecules $r_d \sim r_W = 1.92$ Å, one can estimate $G_Z \sim 2 \times 10^9$. For definiteness, we will be guided by the negative ion O_2^- in atmospheric air, and for the atmospheric temperature $T = 268$ K at the altitude 3 km for standard atmosphere we have for the density of states of the negative ion in formula (4.68) $\xi = 2 \times 10^{26}$ cm^{-3}. In addition, the removal of one negative ion leads to the following change of the ion binding energy

$$\varepsilon_{Z+1} = \varepsilon_Z - \frac{Ze^2}{r}, \quad (4.70)$$

and for the given microdroplet parameters ($r = 8\,\mu$m, $Z = 27$), we have $Ze^2/r \approx 5$ meV and $Ze^2/rT \approx 0.2$.

It should be noted that this description assumes the presence of active centers on the microdroplet surface and the equilibrium under consideration corresponds to the Langmuir isotherm [169] for an equilibrium of the surface with active centers and free absorbed atomic particles. In this case, we reduce this equilibrium to the Saha distribution [170] between the surface and attached negative ions [171]. In this case, the statistical weight G_Z is the number of active centers which are located on the droplet surface and may form a bond with negative ions, and the binding energy of a new negative ion decreases with an increasing number of attached negative ions because of the Coulomb interaction between bound negative ions. But a typical droplet charge is relatively small $Z \ll Z_{\max}$, where $Z_{\max} \sim n^{1/3}$ is the maximum charge of the microdroplet at which the Coulomb interaction for a new joined ion is equalized by the binding energy of this ion with the microdroplet surface. Under the considered conditions, $Z_{\max} \sim 3000$, i.e. the above criterion $Z \ll Z_{\max}$ holds true.

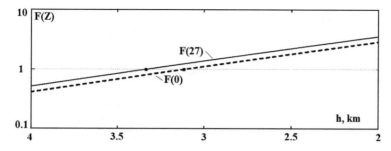

Fig. 4.16   Altitude dependence for the ratio of probabilities $F(Z)$ that a microdroplet has charges $-Z$ and $-Z-1$ correspondingly at the ion number density $N_- = 1\,\mathrm{cm}^{-3}$ for standard atmosphere. The temperature dependence of the quantity $F(Z)$ is taken according to formula (4.68) related to thermodynamic equilibrium (4.67). The binding energy of a negative ion $B^- = O_2^-$ with a neutral water microdroplet is taken as $\varepsilon_0 = 0.9\,\mathrm{eV}$

Figure 4.16 contains the ratio (4.68) as a function of the temperature at some binding energies, if we take the number density of negative ions $N_- = 1\,\mathrm{cm}^{-3}$. This corresponds to the number density of light ions formed under the action of cosmic rays at the first stage of their evolution, before a coat consisting of water and admixture molecules joins with them. Figure 4.16 demonstrates the character of liberation of water microdroplets from a charge in the course of droplet descent in the atmosphere. In the case $F(Z) = 1$, the probabilities for a water microdroplet to have negative charges Z and $Z + 1$ are equal, i.e. this corresponds to the maximum of the charge distribution function of microdroplets. As is seen, when a water microdroplet descends, it liberates from charges if the thermodynamic equilibrium under consideration is established fast.

The above case is convenient for the demonstration of the character of charge release by water microdroplets, though other channels of chemical equilibrium for a chemical bond of a negative ion with the microdroplet surface are possible. But for the formation of such a bond, a fast transition is required between the bound and free states of the negative ion. In order to analyze the possibility of the chemical equilibrium involving the water microdroplet and negative ion, we estimate a typical time of the overbarrier electron transition. Then this time may be estimated according to the Boltzmann formula

$$t_b = \tau_o \cdot \exp(E_b/T), \qquad (4.71)$$

where $\tau_o = 2.4 \times 10^{-17}\,\mathrm{s}$ is a characteristic atomic time constructed on the basis of atomic units, and E_b is the transition energy. Figure 4.17 gives the temperature dependence for an electron transition time on the temperature. Because typical times of variation of atmospheric parameters for various processes exceed minutes, from this one can conclude that the equilibrium is supported for transitions with the energy change below 1 eV.

This analysis allows one to understand the character of microdroplet discharging. In the course of the descending of water microdroplets in the atmosphere, they pass

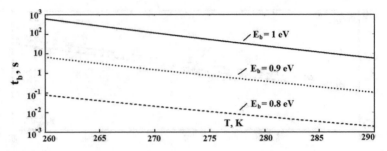

Fig. 4.17 Temperature dependence of the rate of an electron transition for some transition energies

through space regions with an increasing temperature. When this temperature reaches a certain value, an equilibrium is established between bound and free electron states through a certain time, and water microdroplets lose their charge. Then the growth of microdroplets resulting from the conversion of neutral microdroplets in rain drops proceeds fast. Conditions for this chain of processes are fulfilled in a warm season.

References

1. V.F. Hess, Phys. Zs. **113**, 1084 (1912)
2. C.T.R. Wilson, Nature **62**, 149 (1900)
3. J. Elster, H. Geitel, Phys. Zs. **2**, 560–590 (1901)
4. C.T.R. Wilson, Proc. Roy. Soc. **A68**, 151 (1901)
5. J. Elster, H. Geitel, Phys. Zs. **3**, 574 (1902)
6. E. Rutherford, H.L. Cooke, Phys. Rev. **16**, 183 (1903)
7. C.T.R. Wilson, Nature **68**, 102 (1903)
8. J. Elster, H. Geitel, Phys. Zs. **4**, 526 (1903)
9. J.C. McLennan, E.F. Burton, Phys. Zs. **4**, 553 (1903)
10. T. Wulf, Phys. Zeitschrschift **8**(251), 253 (1909)
11. C.T.R. Wilson, Proc. Roy. Soc. **37A**, 32D (1925)
12. R.A. Mullican, Proc. Nat. Acad. Sci. **12**, 48 (1926)
13. R.A. Mullican, Science **81**, 211 (1935)
14. R.A. Mullican, *Electrons(- and +), Protons, Photons, Neutrons and Cosmic Rays* (Chicago University Press, Chicago, 1935)
15. B.B. Rossi, *Cosmic Rays* (McGraw-Hill, New York, 1964)
16. A.M. Hillas, *Cosmic Rays* (Pergamon Press, Oxford, 1972)
17. K. Sakurai, *Physics of Solar Cosmic Rays* (Tokyo University Press, Tokyo, 1974)
18. T.K. Gaisser, *Cosmic Rays and Particle Physics* (Cambridge University Press, 1990)
19. R.K.F. Grieder, *Cosmic Rays of Earth: Researcher's Reference Manuel and Data Book* (Elsevier, Amsterdam, 2001)
20. L.I. Miroshnichenko, *Solar Cosmic Rays* (Kluwer, Dordrecht, 2004)
21. V.I. Dorman, *Cosmic Rays in the Earth Atmosphere and Underground* (Kluwer, Dordrecht, 2004)
22. L. Miroshnichenko, *Solar Cosmic Rays* (Fundamental and Applications (Springer, Heidelberg, 2015)
23. C. Störmer, *Polar Aurora* (Claredon Press, Oxford, 1955)
24. D.F. Smart, M.A. Shea, Adv. Space Res. **36**, 2012 (1955)

25. D.J. Cooke, J.E. Humble, M.A. Shea et al., Nuovo Chim. **C14**, 213 (1991)
26. G.A. Bazilevskaya et.al., in *Planetary Atmospheric Electricity* ed. by F. Leblanc et.al. (Heidelberg, Springer, 2008), p. 149
27. J.A. Simpson, E. Hungerford, Phys. Rev. **77**, 847 (1950)
28. J.A. Simpson, Ann. Rev. Nucl. Part. Sci. **33**, 323 (1983)
29. G.A. Bazilevskaya, M.B. Krainev, V.S. Makhmutov, J. Atmos. Sol.-Ter. Phys. **62**, 1577 (2000)
30. G.A. Bazilevskaya, Space Sci. Rev. **94**, 25 (2000)
31. S. Swordy, D. Mueller, P. Meyer et al., Astrophys. J. **349**, 625 (1990)
32. C. Amster et al., Phys. Lett. **667B**, 1 (2008)
33. A. Erediato, *Cosmic-rays Physics* (2008)
34. https://en.wikipedia.org/wiki/tevatron
35. https://en.wikipedia.org/wiki/Large-Hadron-Collider
36. B.M. Smirnov, *Microphysics of Atmospheric Phenomena* (Springer Atmospheric Series, Switzerland, 2017)
37. L.D. Landau, E.M. Lifshitz, *Quantum Mechanics* (Pergamon Press, Oxford, 1965)
38. H.V. Neher, J. Geophys. Res. **72**, 1527 (1967)
39. H.V. Neher, J. Geophys. Res. **76**, 1637 (1971)
40. G. Plotzer, Zs. Phys. **102**, 23 (1936)
41. M. Nicolet, Planet. Space Sci. **23**, 637 (1975)
42. A. Kryvolutsky et al., Phys. Chem. Earth **27**, 471 (2002)
43. M.A. Ruderman, J.W. Chamberlain, Planet Space Sci. **23**, 247 (1975)
44. A. Hirsikko et al., Boreal Environ. Res. **12**, 265 (2007)
45. K. Nagaraja et al., Radiat. Meas. J. Atm. Sol-Terr. Phys. **68**, 757 (2006)
46. A. Hensen, J.C.H. van der Haage, J. Geophys. Res. **99**, 10693 (1994)
47. R.G. Harrison, H. Tammet, in *Planetary Atmospheric Electricity* ed. by F. Leblanc et al. (Heidelberg, Springer, 2008), p. 107
48. A.M. MacLeod, J.M. Reid, Proc. Phys. Soc. **87**, 437 (1966)
49. J.M. Blatt, V.F. Weiskopf, *Theoretical Nuclear Physics* (Wiley, New York, 1952)
50. V.F. Weiskopf, Rev. Mod. Phys. **29**, 174 (1950)
51. N.A. Jelley, *Fundamentals of Nuclear Physics* (Cambridge University Press, Cambridge, 1990)
52. G.R. Sattler, *Introduction to Nuclear Reactions* (Oxford University Press, New York, 1990)
53. J.S. Lilley, *Nuclear Physics* (Chichester, Wiley, 2001)
54. I. Morrison, *Introduction to Astronomy and Cosmology* (Wiley, Chichester, 2008)
55. R.L. Murray, K.E. Holbert, *Nuclear Energy* (Elsevier, Oxford, 2015)
56. S.I. Akasofu, S. Chapman, *Solar-terrestrial Physics* (Claredon Press, Oxford, 1972)
57. Z. Svestska, *Solar Flares* (Reidel Publishing Company, Dordrecht, 1976)
58. W. Hershel, Phil. Trans. Roy. Soc. **91**, 265 (1801)
59. http://en.wikipedia.org/wiki/corona
60. M.A. Schwanden, *Physics of the Solar Corona* (Praxis Publishing, Chichester, UK, 2005)
61. L. Golub, J.M. Pasachoff, *The Solar Corona* (Cambridge University Press, Cambridge, 2010)
62. R.K. Janev, L.P. Presnyakov, V.P. Shevelko, *Physics of Highly Charged Ions* (Springer, Berlin, 1985)
63. B.M. Smirnov, *Physics of Atoms and Ions* (Springer, New York, 2003)
64. Y.I. Grineva, V.I. Karev, V.V. Korneev et al., Sol. Phys. **29**, 441 (1973)
65. I.I. Sobelman. *Atomic Spectra and Radiative Transitions* (Springer, Berlin, 1979). **Charging of particles**
66. U.S. Standard Atmosphere, (Washington, U.S. Government Printing Office, 1976)
67. R.P. Feynman, R.B. Leighton, M. Sands, *The Feynman Lectures of Physics*, vol. 2 (Addison-Wesley, Reading, 1964)
68. B.M. Smirnov, *Nanoclusters and Microparticles in Gases and Vapors* (DeGruyter, Berlin, 2012)
69. B.M. Smirnov, *Clusters and Small Particles in Gases and Plasmas* (Springer NY, New York, 1999)

70. B.M. Smirnov, Phys. Usp. **43**, 453 (2000)
71. A. Einstein, Ann. Phys. **17**, 549 (1905)
72. A. Einstein, Ann. Phys. **19**, 371 (1906)
73. A. Einstein, Zs.für Electrochem. **14**, 235 (1908)
74. N.A. Fuchs, *Evaporation and Growth of Drops in a Gas (Moscow* (Izd, AN SSSR, 1958). in Russian
75. P. Langevin, Ann. Chem. Phys. **8**, 245 (1905)
76. J. Bricard, in *Problems of Atmospheric and Space Electricity*, ed. by C.C. Coronity (Amsterdam, Elsevier, 1965), p. 82
77. P. Arendt, H. Kallmann, Zs. Phys. **35**, 421 (1926)
78. E.J. Workman, S.E. Reynolds, Phys. Rev. **78**, 254 (1950)
79. G.M. Caranti, A.J. Illingworth, Nature **284**, 44 (1980)
80. J.M. Caranti, A.J. Illingworth, S.J. Marsh, J. Geophys. Res. **90D**, 6041 (1985)
81. J. Latham, Quart. J. Roy. Meteor. Soc. **89**, 265 (1963)
82. J. Hallett, C.R.R. Saunders, J. Atmos. Sci. **36**, 2230 (1979)
83. J.P. Rydock, E.R. Williams, Quart. J. Roy. Meteor. Soc. **117**, 409 (1991)
84. R.L. Ives, J. Franklin Inst. **226**, 691 (1938)
85. I.M. Imyanitov, E.V. Chubarina. *Electricity of the Free Atmosphere* (Israel program for Scientific Translations, Jerusalem, 1967)
86. I.M. Imyanitov, *Electrization of Flights in Clouds and Precipitation* (Gidrometeoizdat, Leningrad, 1970). in Russian
87. S.E. Reynolds, M. Brook, M.F. Gourley, J. Meteorol. **14**, 426 (1957)
88. B. Vonnegut, Bul. Am. Met. Soc. **34**, 378 (1953)
89. V.F. Petrenko, I.A. Ryzhkin, J. Phys. Chem. **101B**, 6285 (1997)
90. V.F. Petrenko, R.W. Whitworth, *Physics of Ice* (Oxford University Press, Oxford, 1999)
91. Y. Dong, J. Yallett, J. Geophys. Res. **97**, 20361 (1992)
92. J.G. Dash, B.L. Mason, J.S. Wettlaufer, J. Geophys. Res. **106**, 20395 (2001)
93. J. Nelson, M. Baker, Atmos. Chem. Phys. Discuss. **3**, 41 (2003)
94. B.J. Mason, *The Physics of Clouds* (Oxford University Press, Oxford, 2010)
95. J. Latham, Quart. J. Roy. Meteor. Soc. **107**, 277 (1981)
96. E.R. Javaratne, C.P.R. Saunders, J. Hallett. Quar. J. Roy. Met. Soc. **109**, 609 (1983)
97. B.L. Mason, J.G. Dash, J. Geophys. Res. **105**, 10185 (2000)
98. P. Berdeklis, R. List, J. Atmosph. Sci. **58**, 2751 (2001)
99. E.R. Williams, R. Zhang, J. Rydock, J. Atmosph. Sci. **48**, 2195 (1991)
100. J.P. Kuettner, Z. Levin, J. Atmos. Sci. **38**, 2470 (1981)
101. C.R.R. Saunders, J. Appl. Meteor. **32**, 642 (1993)
102. B.J.P. Marshall, J. Latham, C.R.R. Saunders, Quart. J. Roy. Meteor. Soc. **104**, 163 (1978)
103. V.F. Petrenko, S.C. Colbeck, J. Appl. Phys. **77**, 4518 (1995)
104. C.R.R. Saunders et al., Atmos. Res. **58**, 187 (2001)
105. W.D. Keith, C.P.R. Saunders, Atmosph. Res. **25**, 445 (1990)
106. C.P.R. Saunders, in *Planetary Atmospheric Electricity*, ed. by F. Leblanc, et al. (Heidelberg, Springer, 2008), p. 335
107. B.M. Smirnov, Phys. Usp. **57**, 1041 (2014)
108. E.E. Avila, M.B. Baker, E.R. Jayaratne, J. Latham, C.P.R. Saunders, Quart. J. Roy. Meteor. Soc. **113**, 1669 (1999)
109. T. Takahashi, J. Atmos. Sci. **35**, 1536 (1978)
110. R.G. Pereyra, E.E. Avila, N.E. Castellano, C.P.R. Saunders, J. Geophys. Res. **105**, 20803 (2000)
111. C.P.R. Saunders, H. Bax-Norman, C. Emersic, E.E. Avila, N.E. Castellano, Quart. J. Roy. Meteor. Soc. **132**, 2653 (2006)
112. B.M. Smirnov, *Cluster Processes in Gases and Plasmas* (Wiley, Berlin, 2010)
113. J. Latham, I.M. Stromberg, The thunder cloud, in *Lightning*, ed. by R.H. Golde (Academic Press, London, 1977), p. 99

114. H.W. Ellis, R.Y. Pai, E.W. McDaniel, E.A. Mason, L.A. Viehland, Atomic Data Nucl. Data Tabl. **17**, 177 (1976)
115. H.W. Ellis, E.W. McDaniel, D.L. Albritton, L.A. Viehland, S.L. Lin, E.A. Mason, Atomic Data Nucl. Data Tabl. **22**, 179 (1978)
116. H.W. Ellis, M.G. Trackston, E.W. McDaniel, E.A. Mason, Atomic Data and Nucl. Data Tabl. **31**, 113 (1984)
117. L.A. Viehland, E.A. Mason, Atom. Data Nucl. Data Tabl. **60**, 37 (1995)
118. F.L. Eisele, P.H. McMurry. Philos. Trans. Roy. Soc. **352 B**, 191–201 (1997)
119. F. Yu, R.P. Turco. J. Geophys. Res. **106 D**, 4797 (2001)
120. K.D. Froyd, E.R. Lovejoy, J. Phys. Chem. **107A**, 9800 (2003)
121. K.D. Froyd, E.R. Lovejoy, J. Phys. Chem. **107A**, 9812 (2003)
122. J. Curtius, E.R. Lovejoy, K.D. Froyd, Space Sci. Rev. **125**, 159 (2006)
123. M.B. Enghoff, J.O.P. Pedersen, T. Bondo et al., J. Phys. Chem. **112**, 10305 (2008)
124. F. Arnold, Space Sci. Rev. **137**, 225 (2008)
125. J. Kazil, R.G. Harrison, E.R. Lovejoy, Space Sci. Rev. **137**, 241 (2008)
126. H. Heitmann, F. Arnold, Composition measurements of tropospheric ions. Nature **306**, 747 (1983)
127. M.D. Perkins, F.L. Eisele, First mass spectrometric measurements of atmospheric ions at ground level. J. Geophys. Res. **89**, 9649 (1984)
128. F.L. Eisele, Identification of tropospheric ions. J. Geophys. Res. -Atmos. **91**, 7897 (1986)
129. F.L. Eisele, J. Geophys. Res. **94**, 2183 (1989)
130. F.L. Eisele, J. Geophys. Res. **94**, 6309 (1989)
131. H. Tammet, J. Aerosol Sci. **26**, 459 (1995)
132. G. Beig, G. Brasseur, J. Geophys. Res. **105**(22), 671 (2000)
133. R.G. Harrison, K.S. Carslaw, Rev. Geophys. **41**, 1012 (2003)
134. U. Horrak, J. Salm, H. Tammet, J. Geoph. Res.: Atmosph. **108**, 62 (2003)
135. M. Kulmala, L. Laakso, K.E.J. Lehtinen et al., Atmosph. Chem. Phys. **4**, 2553 (2004)
136. H. Tammet, Atmosph. Res. **82**, 523 (2006)
137. R.G. Harrison, K.L., Aplin, Atmosph. Res. **85**, 199 (2007)
138. A. Hirsikko, T. Nieminen, S. Gagn et al., Atmosph. Chem. Phys. **11**, 767 (2011)
139. L.D. Landau, E.M. Lifshits, *Fluid Mechanics* (Pergamon Press, London, 1959)
140. J.J. Thomson, Philos. Mag. **47**, 334 (1924)
141. B.M. Smirnov, *Physics of Ionized Gases* (Wiley, New York, 2001)
142. J. Sayers, Proc. Roy. Soc. **A169**, 83 (1938)
143. W. Mächler, Zs. Phys. **104**, 1 (1936)
144. H. Israël, *Atmospheric Electricity* (Keter Press Binding, Jerusalem, 1973)
145. B.J. Mason, *The Physics of Clouds* (Claredon Press, Oxford, 1971)
146. https://windexchange.energy.gov/maps-data
147. https://sciencing.com/average-daily-wind-speed
148. https://www.currentresults.com/Weather/US/wind-speed-city-annual
149. D. Malan, Ann. Geophys. **8**, 385 (1952)
150. D. Malan, *Physics of Lightning* (The English University Press Ltd., London, 1963)
151. https://en.wikipedia.org/wiki/Thunder
152. http://www.waterencyclopedia.com/Po-Re/Precipitation-and-Clouds-Formation
153. B.F.J. Schonland, *Atmospheric Electricity* (Methuen, London, 1932)
154. J.A. Chalmers, *Atmospheric Electricity* (Claredon Press, Oxford, 1949)
155. Yal Frenkel, *Theory of Phenomenon of Atmospheric Electricity* (GITTL, Leningrad, 1949). in Russian
156. B.F.J. Schonland, *Atmospheric Electricity* (Methuen, London, 1953)
157. B.F.J. Schonland, *The Lightning Discharge. Handbuch der Physik*, vol. 22 (Springer, 1956), p. 576
158. H. Israël, *Atmospheric Electricity. vol. 1. Fundamentals, Conductivity.* (Academische Verlagsgesellschaft, Leipzig, 1957)

159. H. Israël, *Atmospheric Electricity*, vol. 2 (Charges, Currents, Academische Verlagsgesellschaft, Leipzig, Fields, 1961)
160. J.A. Chalmers, *Atmospheric Electricity* (Pergamon Press, Oxford, 1967)
161. K. Berger, The earth flash, in *Lightning*, ed. by R.H. Golde. (Academic Press, London, 1977), p. 119
162. M.A. Uman, *Lightning* (McGrow Hill, New York, 1969)
163. M.A. Uman, *About Lightning* (Dover, New York, 1986)
164. M.A. Uman, *The Lightning Discharge* (Academic Press, New York, 1987)
165. V.A. Rakov, M.A. Uman, *Lightning, Physics and Effects* (Cambridge University Press, Cambridge, 2003)
166. V.A. Rakov, *Fundamental of Lightning* (Cambridge University Press, Cambridge, 2016)
167. N.B. Vargaftic, *Tables of Thermophysical Properties of Liquids and Gases* (Halsted Press, New York, 1975)
168. B.M. Smirnov, *Reference Data on Atomic Physics and Atomic Processes* (Springer, Heidelberg, 2008)
169. C. Kittel, *Thermal Physics* (Wiley, New York, 1970)
170. M.N. Saha, Proc. Roy. Soc. **99 A**, 135 (1921)
171. E. Illenberger, B.M. Smirnov, Phys. Usp. **41**, 651 (1998)

Chapter 5
Processes of Atmospheric Electricity

Abstract Dividing the atmosphere in two parts, where the main part corresponds to a clear sky, one can represent electric processes in this atmospheric part as the global electric circuit. In this consideration, the Earth and its atmosphere may be represented as a spherical capacitor, one electrode of which is the Earth's surface, and the other one is an atmospheric layer with high conductivity. The main contribution to the atmospheric resistance follows from the troposphere. The other atmospheric part with cumulus clouds which cover a few percent of the Earth's surface is responsible for charging the Earth. Evolution of study of atmospheric electricity gives the understanding of this problem at that time and leads to the contemporary state of this atmospheric science. Lightning is the most brightest phenomenon of atmospheric electricity and is studied in detail both on the basis of observations and from theoretical analysis. Here, it is analyzed as an ionization wave which creates the channel of dissociative air and propagates along it. Properties of this conductive lightning channel and its evolution are analyzed. The criterion is given for repetition of lightning flashes.

5.1 Global Electric Circuit

5.1.1 Earth as Electrical System

Let us represent average electric parameters of the Earth and its atmosphere. This problem is analyzed in detail in review [1]. The Earth contains a negative charge [2–4] that is equal on average $Q = R_\oplus^2 = 5.8 \times 10^5$ C [5, 6]. This charge creates an electric field of strength $E_o = Q/R_\oplus^2 = 130$ V/m at the Earth's surface, where

© The Editor(s) (if applicable) and The Author(s), under exclusive license
to Springer Nature Switzerland AG 2020
B. M. Smirnov, *Global Atmospheric Phenomena Involving Water*,
Springer Atmospheric Sciences, https://doi.org/10.1007/978-3-030-58039-1_5

$R_\oplus = 6370\,\text{km}$ is the Earth's radius. This allows one to model the troposphere as an electric system by a spherical condenser with relatively high conductivity at its electrodes. The electric potential between the electrodes is $U_o = 240 - 300\,\text{kV}$ [1, 7, 8].

Next, according to the observational data, the average density of the current that discharges the Earth is $2.4\,\text{pA/m}^2$ over the land on average, and over the oceans it is on average of $3.7\,\text{pA/m}^2$ [5]. This corresponds to the total discharge current of the Earth passing through its atmosphere, equal to $I = 1700\,\text{A}$ [5, 9]. This gives for the average conductivity Σ of a quiet atmosphere

$$\Sigma = \frac{I}{E_o s} = 2.6 \times 10^{-14}\frac{S}{m} = 2.3 \times 10^{-4}\text{s}^{-1}$$

where S is the unit Siemens, $s = 5.1 \times 10^{14}\text{m}^2$ is the area of the Earth's surface. In addition, if we take the mobility of atmospheric ions to be $2\,\text{cm}^2/(\text{V} \cdot \text{s})$, as they have on the stage of their formation [10], one can find the average time of Earth discharging as $\tau = Q/I \approx 6\,\text{min}$.

Continuing to exploit the global electric model of the Earth atmosphere, we obtain that the atmosphere resistance R_a is equal

$$R_a = \frac{U_o}{I} = 150\,\Omega \tag{5.1}$$

Figure 5.1 gives the distribution of the average atmosphere resistance over altitudes [7, 8]. Indeed, the atmosphere resistance is given by

$$R_a = \int_0^\infty \frac{dh}{\sigma(h)}, \tag{5.2}$$

Fig. 5.1 Contribution to the atmosphere resistance from atmospheric layers up to indicated altitudes.
1—according to the analysis [7, 8], 2—the model with $\sigma = \text{const at } h < h_o$

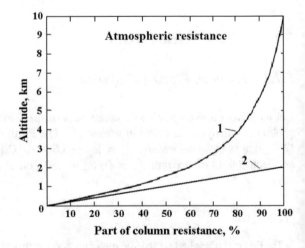

Table 5.1 Typical electric parameters of the global atmosphere during the fair weather [18] (U is the electric potential of the upper atmosphere with respect to the Earth's surface, E is the electric field strength near the Earth's surface, h is the altitude, i is the current density through the atmosphere, I is the total current, Ω is the atmosphere resistance, ρ is the resistance per unit area of an atmospheric column)

Parameter	Value
U, kV	250
dE/dh, V/m	120
i, pA/m^2	3
I	1800
R_a, Ω	230
ρ_a, pΩ/m^2	120

where the total conductivity $\sigma(h)$ is averaged over the Earth surface. As is seen, the main contribution to this resistance follows from layers with a small conductivity. We give in Fig. 5.1 the above simple approximation where the electric field is concentrated at low altitudes $h \leq h_o$ with the constant conductivity $\sigma(h) = $ const. In this approximation on the basis of the ionosphere voltage $U_o = 250$ kV and the electric field strength $E_o = 130$ V/m, we have $h_o = U_o/E = 2$ km. This approximation is represented in Fig. 5.1 by curve 2.

Note that these values were obtained a century ago and are the object of monographs [5, 9, 11–17]. In addition, we give in Table 5.1 values of the above average parameters according to [18]. These values may differ slightly from those used above.

It is required a certain caution in description of the Earth and its atmosphere as a spherical capacitor because of a field nonuniformity both in horizontal and vertical directions [19]. This relates to the space distribution of the atmospheric conductivity [20] and the electric field strength [21]. Nevertheless, the above simple model for description of the electric phenomena allows one to understand electric properties of the troposphere.

5.1.2 Electric Machine of Troposphere

A simple model for the Earth's atmosphere in studying its electrical properties is the spherical capacitor, whose lower electrode is the Earth's surface, and the upper electrode is the Earth's layer of high conductivity, usually, it is the ionosphere. Note that though the ionosphere is believed as the upper electrode of the global electric circuit is the ionosphere, 88% of the ionospheric electric potential falls in the troposphere. This consideration allows one to formulate the concept of a global electric circuit [3, 22–25]. In reality, the possibility to represent some electric aspects of the atmosphere in the form of a global electric circuit follows from different processes of charging the Earth and its discharging. Since this concept exists more a century,

Fig. 5.2 Carnegie
curve—the dependence of
the electric field strength
near the Earth's surface
(Potential Gradient) on the
hour of day (Universal Time)
[28]. The point relates to the
average values, grey region
characterizes the distribution
of data over electric field
strengths

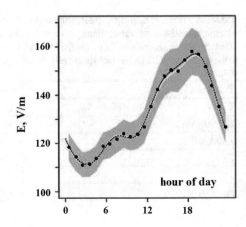

various details of these concepts are investigated (for example, [1, 18, 24, 26, 27]),
and, therefore, we analyze briefly only some of them which can be used for the
analysis of microphysics of ion processes.

Because the global electric circuit is connected with processes of ion formation and
their loss, and conditions for these processes vary, the global atmosphere circuit varies
with variation in the atmosphere parameters. Because these parameters are connected
with solar processes in the atmosphere, parameters of the global atmosphere circuit
may vary both in different time of day and season. This peculiarity is realized in the
Carnegie curve that is the dependence of the electric field strength in the atmosphere
near the Earth's surface on an hour of day in a fair weather. The above name followed
from ship cruises organized by the Carnegie Institution of Washington. According to
measurements, minimal electric field strengths are observed approximately at 03UT
(3 h of Universal Time), and the maximum electric field strength refers to 19 UT.
Figure 5.2 contains a typical Carnegie curve which uses observed data from various
geographic points. This testifies about the universality of the Carnegie curve and
allows one to understand the complexity of atmospheric electricity one century ago
[29–33].

One can explain this from a general standpoint. In the global atmosphere, the
charge separation results from falling of charged water microdroplets located mostly
in cumulus clouds. Since water microdroplets are absent in a fair atmosphere,
aerosols, i.e. atmospheric particles, play a role in these processes. At land these
aerosols are created by trees or forests [34–37]. At oceans, such aerosols are a result
of water evaporation from the surface. This mechanism contradicts partially to a general character of creation of atmospheric electricity due to falling down of charged
water microdroplets, and the latter process leads to charge separation in the atmosphere. The main contribution to this separation follows from atmospheric regions
containing cumulus clouds with a high number density of water microdrops. In any
case, a typical size of aerosols which can lead to charge separation is large and
satisfies the criterion (4.33).

One can overcome the above contradiction partially due to a high conductivity of the Earth's surface and ionosphere compared to that of the troposphere and low atmosphere. Indeed, the average current density for discharging of our planet is equal to $i = 2.4\,pA/m^2$ over land and to $i = 3.7\,pA/m^2$ over oceans [5, 38]. This corresponds the average conductivity of the Earth's atmosphere $\sigma = i/E = 2.5 \times 10^{-14}\,S/m$, where we take $E = 130\,V/m$, and the conductivity unit Siemens is $S = 1/\Omega$ [39]. The conductivity of the underground is higher than that for the atmosphere and ranges from $10^{-7}\,S/m$ for rocks up to $10^{-2}\,S/m$ for clay and limestone soil [24, 40]. The average conductivity of oceans is $3\,S/m$ [41] and depends on the presence of sodium and potassium salts in water. The conductivity of the lower ionosphere at the altitude of $80\,km$ is of the order of $10^{-7}\,S/m$ [7, 8] and exceeds that of the troposphere ($\sim 10^{-14}\,S/m$). Indeed, the ionosphere conductivity is created by electrons, the electron number density of the ionosphere exceeds that of tropospheric ions and the number density of molecules for scattering of charged atomic particles is lower in the ionosphere than that in the troposphere. Thus, the conditions of the global electric circuit model may be provided if separation of atmospheric charges is created only in cloud regions. Then charges of the crust and ionosphere propagate over a globe that leads to creation of identical electric fields both in fair atmospheric regions and those with cumulus clouds.

The Carnegie effect discovered over oceans results from different concentrations of atmospheric admixtures which influence on ion growth and recombination in the atmosphere. Finally, this leads to different electric fields over the oceans depending on a day time. Though these ions were created in the same manner under the action of cosmic rays, weak processes in the atmosphere, including those as a result of human activity, may influence on the atmosphere electric state. In particular, the presence of radioactivity in the atmosphere leads to additional ionization that changes electric parameters of the atmosphere. An accident at the Chernobyl nuclear power plant in April 1986 caused such variation. Observations of thunderstorm events several months after this accident in Sweden and its analysis confirm this statement [42, 43]. Under the action of enhanced ionization of atmospheric air, the rate of lightning occurring is increased. Hence, the Chernobyl accident led to intensification of thunderstorms during several months enough far from the place of this accident.

We have that the concept of the global electric circuit exists almost one century and is studied in detail. In particular, some aspects of this phenomenon were investigated in [1, 27, 44–50]. Against this background, the above analysis based on simple models gives a simplified description of this as an element of the circulation of atmospheric electric currents. In addition to the above consideration of electric current through the clear atmosphere, we model it by a spherical capacitor in which electrodes are the Earth's surface and ionosphere. If such electrodes absorb weakly an electromagnetic wave, one can consider this system as a waveguide in which electromagnetic wave may be propagated. The standing electromagnetic wave in this waveguide is called the Schumann resonance [51–53]. In reality, such an electromagnetic wave is excited by lightning and propagates over the world inside the above waveguide. One can determine the frequency ω_S of the basic resonance at the equator as [54] $\omega_S = L/c \approx 7.5$ Hz, where $L \approx 40\ 000$ km is the equator length, and c is the light speed. Because the Earth is almost round, the frequencies of propagation in different directions are nearby. In reality both the basic frequency ω_S of the Schumann resonance and frequencies of overtones may be changed slightly depending on atmospheric parameters [55, 56]. Therefore, Schumann resonances may be used as an indicator of the atmosphere state, in particular, as a global tropical thermometer [57].

5.1.3 Study of Atmospheric Electricity and Lightning

Lightning is the bright phenomenon of atmospheric electricity. The above analysis shows that lightning is an instability which develops in the course of falling of charged water microdroplets if they occur in regions with a high temperature where a release of bound electrons proceeds from microdroplets. Lightning as the atmosphere breakdown is energetically profitable because cumulus clouds acquire a high electric potential with respect to the ground. Nevertheless, this electric breakdown does not mean charge transport between the clouds and ground. In this case, the charge transport proceeds in the ground because of a higher conductivity compared with that of the atmosphere. This charge transport decreases the potential difference between clouds and ground.

From the standpoint of the Earth's charging, lightning is not dominate channel, because contribution to the Earth's charging is approximately 20%. But lightning is the most remarkable atmospheric phenomenon, and, therefore, it is studied in more detail compared with other atmospheric processes. In addition, this is an electric process which has an analogy with laboratory electric discharges, and latter is important for understanding the atmospheric processes. Hence, below we consider some stages of development of sciences which promote to understand the nature of lightning as atmospheric electric breakdown, though the choice of these stages has an arbitrary character.

Let us start this analysis from 1705, when the English scientific Hauksbee [58, 59] has made a step to scientific instrumentation and experimentation. He constructed the first powerful electrostatic generator that allowed him to produce electrical discharges in air which are accompanied by air glowing. This was the step to investigate electric gas discharges as the passage of electric currents through a gas. Finally, this allows one to understand the nature of this phenomenon. On this way, Stephen Gray (England) in 1731 introduced the conductivity as the gas property which characterized the pass of electricity through it [59, 60].

In 1734 1734 C.F. de Cisternay Dufay (France) [61] showed the existence of two types of electrification, which are different for dielectric and metals. From this, it follows two types of electrical flows. C.F.C.Dufay also discovered a heightened conductivity of air near hot objects that was an evidence of the air conductivity at high temperatures. In 1745, E.J. Von Kleist (Germany) and P. Van Musschenbroek (Netherlands) independently invented an electric capacitor [62] which was called the

"Leyden jar" [63]. The Leyden jar consists of a glass jar covered inside and outside with metal foils and two such hemispheres are separated by a rubber stopper. If the jar is charged with a help of an electric machine, the force of two horses did not allow to separate these hemispheres.

Experimental investigations of 1752 by American scientist Benjamin Franklin [59, 64] on the basis of the above understanding proves the electrical nature of lightning and allows one to consider it as an electrical current propagated through the atmosphere. He proves that the electric conductivity results from transfer of electrical charges. An important result of Franklin investigations was the conclusion that there are two types of carriers of the electric current. Investigations of electrolysis by Michael Faraday (England) in 1833–1834 allowed one to determine parameters of electricity carriers connecting the electricity amount q which passes through the electrolyte, the mass m of a transferred substance which is extracted at an electrode, and A is the atomic weight of electricity carriers. Finally, the Faraday laws for connection between these parameters may be represented in the form [65]

$$q = F \cdot \frac{m}{A} \tag{5.3}$$

Here, q is the electric charge passed through the electrolyte, m is the mass of a transferred substance, A is its atomic weight, i.e. m/A is the mass of a transported substance expressed in moles. The proportionality coefficient F or the Faraday constant is equal to $F = 96485\,C/mole$. In addition to this, Faraday introduced in physics new terms for quantities which are responsible for the passage of electricity through a matter. There are among these terms "ion", "anode" (a path down, in Greek), and "cathode" (a path up, in Greek).

As a continuation of study of elementary electricity carriers, one can consider investigation of gas discharge of low pressure. In the course of study of such gas discharge, improved vacuum tubes were elaborated by Julius Pluecker in 1857 [66] who observed a light on the cathode surface, if the tube is located in a magnetic field. Subsequently, its student and collaborator Wilhelm Hittorf [67] shows that this light called as cathode rays was deflected by a magnetic field. These cathode rays were studied as the flux of elementary electricity carriers, electrons. Finally, J.J.Thomson [68] determined in 1897 the charge-to-mass ratio for electrons as elementary electricity carriers in crossed electric and magnetic fields that occurred large compared to that for atoms. As a result, two types of electricity carriers were discovered, electrons and ions, and transport of these charged particles determines the electric processes in the atmosphere.

Along with the above universal concepts of physics, the understanding of atmospheric electric phenomena is based also on their analogy with processes of electric breakdown of gases [69–74]. For electric breakdown of gases, it is necessary that the electric field strength between two electrodes exceeds the breakdown field which for dry air at atmospheric pressure is $E_o = 30\,kV/cm$. In this case, multiplication of electrons in a gas creates an electric current, and the electric field results from displacement of electrons which screen partially the electric field between electrodes.

As a result, the equilibrium is established between the current which discharges the electrodes and the charge located at electrodes or the electric potential between them. But the passage of the electric current through a gas has a self-consisted character and creates a specific medium—an ionized gas in the form of a spark [75–77].

Lightning differs from an electric spark because a typical electric field strength in the atmosphere during a thunderstorm weather is 200 V/cm that is less two orders of magnitude than the breakdown one. Hence, lightning is an ionization wave and, therefore, is akin to a streamer form of gas discharge. Such development of gas discharge in dense gases was studied for the streamer form of gas discharge. Principles of the streamer form of gas discharge were elaborated by Raether [71, 78–85], Loeb [69, 75, 86–89] and Mick [70, 90–93], starting from thirties of twentieth century. The concept of the streamer as an ionization wave consists in redistribution of the electric potential in the course of its propagation in a space between two electrodes such that the electric field strength increases near the front of the ionization wave and exceeds there the breakdown one. This leads to intense gas ionization near the front of the ionization wave. Photoionization of a gas is of important element of this wave, so that a photon emitted near the front of the ionization wave is absorbed at some distances from it, and formed electrons ionize the gas in this region.

The above character of propagation of lightning as an ionization wave leads to a high velocity of the wave compared with a typical electron velocity. It should be noted that according to the ionization equilibrium, the electric field strength exceeds the threshold one for air breakdown in some regions. This is realized also in the cathode region of the Townsend regime of gas discharge [94–97]. The choice between the Townsend regime of gas discharge and regime through propagation of the ionization wave is determined by the degree of the overvoltage [98, 99]. It should be noted that though the character of propagation of lightning resembles that of a streamer, due to a higher electric current of lightning, properties of an ionized gas in lightning differ from those for a streamer.

Along with understanding the fundamental and specific electric processes in lightning, the important role in description of this phenomenon follows from measurements of its parameters that is facilitated due to a strong intensity of this phenomenon. Dynamics of lightning follows from its photographies with the Bois chamber which basis is the Kerr cell as a shutter for its action [100]. The Kerr shutter provides a switching time up to $10ns$ that allows one to measure propagation of light for distances of a meter range. The velocity of lightning propagation is one-two order of magnitude below the light speed.

5.1.4 Global Electric Atmospheric Processes

We now summarize the above analysis of some electric processes in the atmosphere. These processes result in circulation of atmospheric current which is charging and discharging the Earth. The charging process results in gravitation falling of charged water microdroplets, and the discharging process follows from electric currents which are created by negative and positive atmospheric ions formed under the action of cosmic rays. Processes of charging of the Earth proceed in cumulus clouds in which practically all atmospheric condensed water is concentrated, while the discharging processes take place in the clear-sky atmosphere, i.e. in the atmosphere which is practically free from clouds. Though lightning processes give a contribution to the Earth's charging, we exclude lightning from a global electric scheme of the Earth's atmosphere because this contribution is restricted. Below we repeat principles of the global atmospheric electricity and collect some parameters which characterize global electric currents in the atmosphere.

Using numerical parameters of electric atmospheric properties, we hold to an approach of this book where we take some typical parameters of objects or processes under consideration as their average values. This simplifies the analysis, but its results have the qualitative character and may be considered as estimations. In particular, in this analysis a cumulus cloud is taken as uniform air containing water microdroplets with parameters (3.1), i.e. it includes water microdroplets of a radius $r = 8\,\mu$m and of the number density $N_d = 10^3$cm^{-3}. The charge of an individual microdroplet is $Z = 27$, as it follows from the lifetime of a cumulus cloud [101]. Next, as it was established by Wilson a century ago [2, 3, 22], the Earth is charged negatively such that the average electric field strength at a faint weather is 130 V/m.

The assumption that charging of the Earth as a result of gravitation fall of charged microdroplets is connected with criteria for a droplet size. First, along with the gravitation force—its weight, the electric field acts on the microdroplet. A droplet moves toward the Earth, if the gravitation force dominates, and then its radius satisfies to the criterion (4.33) $r > 0.4\,\mu$m. On the other hand, the ratio of the droplet charge to its mass exceeds this ratio (4.52) for total fluxes of atmospheric charge and water. This accounts for the proportionality between the flux of evaporated water from the Earth's surface and the charging atmospheric current to the Earth charging, because electric atmospheric phenomena are secondary ones with respect to water circulation. In other words, this ratio exceeds the value

$$\frac{dq}{dm} = 1.1 \times 10^{-10}\,\text{C/g}$$

given by formula (4.52). This leads to the following criterion for the droplet radius [10]

$$\frac{r}{Z^{1/3}} < 20\,\mu\text{m}, \tag{5.4}$$

where the ratio of a transported charge to transferred water in the form of micro-droplets is equal 1.1×10^{-10} C/g in accordance with formula (4.52). Droplet parameters under consideration ($r = 8\,\mu$m, $Z = 27$) satisfy to the criterion (5.4). Moreover, they give the charge-to-mass ratio 2×10^{-9} C/g [101, 102] for microdroplets through which the charge transport is realized. From this, it follows that the mass of atmospheric condensed water contained mostly in cumulus clouds is estimated as several percent of the total mass of atmospheric water which is located there in the form of free water molecules.

Next, based on observational data for the electrical properties of the atmosphere, we connect them with the above estimates. Let us assume that the length of the charged cloud significantly exceeds its thickness. The electrical potential of the pre-thunderstorm cumulus cloud is 20–100 MV [103]. For a typical cloud altitude (2–4) km, this corresponds to the observed electric field strength $E = 200$ V/cm. For a relatively thin charged layer, this gives for the surface charge density of the cloud

$$\Sigma = \frac{E}{4\pi} \sim 1 * 10^8 \, e/cm^2 \qquad (5.5)$$

Taking the above parameters of microdroplets, one can estimate the charge density of a cumulus cloud as $\sim 3 \times 10^3$ cm^{-3}. From this, one can estimate the thickness of a uniform cumulus cloud, that is, ~ 30 km. It is necessary to take into account that a cumulus cloud is nonuniform and contains both regions with charged microdroplets and also regions without microdroplets, but with positive ions. Then a thickness of a cumulus cloud that creates a heightened electric field in a thunderstorm weather is estimated as

$$l \sim 100 \, m \qquad (5.6)$$

From this, one can estimate also which area over a ground partakes in thunderstorm. Indeed, the electric charge which passes to a ground as a result of one lightning flash transfers a charge (4–5) C according to measurements [104–106]. According to the charge density (5.5), this charge is gathered from the area over the earth S of the order of $S \sim 30$ km^2. As is seen, clouds of large sizes partake in thunderstorms.

Under these conditions, taking the charge density due to charged water micro-droplets $N_Z = 3 \times 10^4$ e/cm^3 and the falling velocity (2.22) $w = 0.8$ cm/s, one can obtain the current density due to charged water microdroplets

$$i = wN_Z = 40 \, pA/m^2$$

This current of the Earth charging is compensated by the discharging current which is realized in atmospheric regions with rare clouds. For the latter value, we are guided by measurements of the book [5]. According to this, the mean current density through a fine atmosphere is 2.4 pA/m^2 for the land and is 3.7 pA/m^2 over oceans. Comparing electric currents of charging and discharging the Earth, one can estimate that the area over the earth, where cumulus clouds are located, is several percent of the total area of the Earth's surface. This is in accordance with the preceding estimation which follows from the analysis of the atmospheric charge density.

Note that in considering atmospheric condensed water, we restrict by cumulus clouds where the basic part of atmospheric condensed water is located. Correspondingly, electric processes in the atmosphere are connected with cumulus clouds mostly. Other types of clouds are rare compared with cumulus ones and they are not of interest for electric properties of the atmosphere. But they are of importance for optical atmospheric properties, and condensed water of these rare clouds is responsible for emission of the atmosphere toward the Earth and outside, rather the cumulus ones.

We now consider one more principal problem of atmospheric electricity which describes pre-thunderstorm processes. Before a thunderstorm, an upstream of wet warm air rises to cumulus clouds and heats them. As a result, water microdroplets lose the charge and grow fast through the coagulation and gravitation mechanisms. In addition, mixing of air of a cumulus cloud with warm air from the ascented flux leads to a temperature increase as well as to an increase of the concentration of condensed water. Our goal now is to determine the part of injected water in a upstream which is converted into condensed one.

The criterion of water condensation in atmospheric water is formulated in Sect. 3.2.2, where mixing of two air layers (parcels) with different temperatures and moistures is analyzed. Then condensation takes place if the supersaturation degree (3.13) exceeds one. We now have other conditions compared to Sect. 3.2.2 which correspond to parameters of a cumulus cloud where at the beginning the supersaturation degree is one. Addition of atmospheric air from the near-surface layer to a region of a cumulus cloud leads to a partial water condensation of added air. Our goal is to determine which part of added water is transformed in the condensation phase.

In this consideration, we use an analogy of mixing of air parcels in the regime under consideration with that of Sect. 3.2.2. In the beginning, the upper parcel contains n_a air molecules and n_w water molecules, and this is mixed with a parcel of hot wet air which contains n_a' air molecules and n_w' water molecules, where $n_a' \ll n_a$, as well as $n_w' \ll n_w$. For definiteness, we take below an additional parcel as saturated air near the Earth's surface, so that $c_E = n_w/n_a = 1.7\%$, and the upper parcel is located at an altitude $h = 3$ km at the beginning, where the temperature is equal $T = 268 K$, and the concentration of a saturated water vapor is $c_w = 0.43\%$. Next, the condensation process leads to energy extraction of ε_b per water molecule which attaches to a water microdroplet. One can equalize the binding energy ε_b per one water molecule with the value $E_{sat} = 0.48$ eV in formula (3.3).

We consider the regime where a water vapor remains a saturated one after addition of warm wet air. Denoting by ξ the probability that added water molecules transfer in the condensed phase, we obtain on the basis of relation (3.3) and the condition of a saturated water vapor in the course of this mixing, as it follows from the Clausius–Clapeyron relation [107, 108]

$$n_w'(1 - \xi) = \frac{\varepsilon_b \Delta T}{T^2} n_w,$$

where ΔT is an increase of the air temperature as a result of attachment of water molecules to microdroplets. This equation may be rewritten in the form

$$\Delta T = \frac{T^2}{\varepsilon_b} \cdot \frac{c_E}{c_w} \cdot \frac{n_a'}{n_a} \tag{5.7}$$

One can use also the equation of energy balance in this process

$$c_p n_a \Delta T = c_p n_a' (T_E - T) + \varepsilon_b \xi n_w'$$

Here, $c_p = 7/2$ is the thermal capacity of an air molecule, temperatures are measured in energetic units, and we assume that a thermal energy of air molecules is not lost in the course of the transport process. This equation may be transformed to the form

$$\Delta T = (T_E - T) \cdot \frac{n_a'}{n_a} + \frac{\xi \varepsilon_b c_w}{c_p} \cdot \frac{n_a'}{n_a} \qquad (5.8)$$

The set of (5.7) and (5.8) allows one to determine the portion ξ of added water molecules which form the condensed phase. We have for the probability ξ to transfer into the condensed state for a water molecule from the added parcel

$$\xi = \frac{1 - \gamma}{1 + \zeta}; \quad \gamma = \frac{(T_E - T)\varepsilon_b c_w}{T^2 c_E}, \quad \zeta = \left(\frac{\varepsilon_b}{T}\right)^2 \cdot \frac{c_w^2}{c_E c_p} \qquad (5.9)$$

In particular, in the case of mixing of the air flux from the Earth's surface with air of a cumulus cloud at an altitude $h = 3$ km which corresponds to the temperature $T = 268$ K of standard atmosphere, we obtain for parameters of formula (5.9): $\gamma = 0.25$, $\zeta = 0.34$, $\xi = 0.56$.

On the basis of the above formulas, we now analyze the character of the process under consideration. Taking into account that air of a cumulus cloud is saturated, we have that transport of air from the Earth's surface to an altitude of cumulus cloud location leads to the heat release through two channels. The first one results from different temperatures of transported air from the Earth's surface and the cumulus cloud. The other channel follows from transition of a part of water molecules from a flux into condensed phase that leads to heat extraction. Under the action of these processes, the air temperature in the cumulus cloud increases, as well as the number density of free water molecules at a new temperature. If this difference for a new number of free water molecules in the cumulus cloud region exceeds that contained in the air flux, then water molecules from the air flux condensate.

Figure 5.3 represents the portion of the part of atmospheric water which is present in the air flux from the Earth's surface to a cumulus cloud depending on the altitude h of mixing. The temperature T of air in the mixing region and the altitude h of this region for the standard atmosphere model are connected by the relation

$$T = T_E - h\frac{dT}{dh}, \quad T_E = 288K, \quad \frac{dT}{dh} = 6.5K/km$$

Next, the concentration of water molecules for a saturated water vapor in the vicinity with the Earth's surface is equal $c_E = 0.017$. According to formulas (2.3) and (2.5), the water concentration at the saturated water pressure for an atmospheric layer with the temperature T is equal

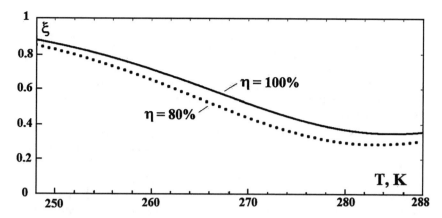

Fig. 5.3 Probability of attachment to water microdroplets for water molecules located in the air flux which is formed in a near-surface region and is mixed with a cumulus cloud of a given temperature. The altitude temperature distribution is taken due to the model of standard atmosphere, η is the moisture of near-surface air

$$c_w = c_E \cdot \exp\left(\frac{E_a}{T_E} - \frac{E_a}{T} - \frac{T_E - T}{T_o}\right), \quad T_o = \Lambda\frac{dT}{dh} = 55K,$$

and the binding energy $E_a = 0.48\text{eV}$ for a water molecule bonded with a micro-droplet is taken from formula (3.3). The probability ξ in formula (5.9) corresponds to water molecules of an air flux which is formed near the Earth's surface relates to the saturated vapor pressure of atmospheric water there, i.e. the moisture is equal $\nu = 100\%$. In the case of any moisture η of the air flux, it is necessary to replace the parameter c_E in formula (5.9) by ηc_E.

5.2 Lightning as a Phenomenon of Atmospheric Electricity

5.2.1 Lightning as a Physical Phenomenon

The most bright phenomenon of atmospheric electricity is lightning as the electric current passage from clouds to the ground. As a phenomenon of atmospheric electricity, lightning leads to charging the Earth. Though the contribution from lightning to charging of the Earth is restricted by approximately 20%, the charging process here is more remarkable because it is concentrated in a small atmosphere region. Therefore, various features of this phenomenon is studied in detail and the results are represented in lightning books [105, 106, 109–114]. Lightning is a complex physical phenomenon in which description cannot follow from simple mathematic models. The analysis of lightning properties is represented in [76, 113, 115–117]. Below on the basis of the above measurements and the analysis, we consider principal aspects of this phenomenon.

In considering lightning as a chain of some processes, one can take the propagation of a red stepwise leader as the first stage of lightning evolution. A stepwise leader is a weakly glowing phenomenon which propagates along a broken line with a typical segment length of (50–100) m. Propagation of a leader may be considered as an ionization wave and finally it creates a conductive channel of lightning. After passage of each segment, the ionization wave stops and then moves with a slight change of the propagation direction. Like to a streamer, this ionization wave is a self-consistent weakly ionized air with a heightened electric field strength at the front of this ionization wave and relatively small electric fields at other parts of the wave. The average velocity of propagation of the stepwise leader with accounting for its stops is of the order of 10^7 cm/s that corresponds to the electron drift velocity in such electric fields. But in the course of wave propagation along a segment the drift velocity is higher by one order of magnitude. When a conductive channel is created, an electric current propagates along it and is accompanied by glowing of the plasma.

The next stage of lightning evolution is the recurrent stroke which proceeds with a velocity up to 5×10^9 cm/s. This velocity corresponds to propagation of an electric signal in conductors. The recurrent stroke is short and lasts a few μs; as a result of this stage, a conductive channel of lightning is formed and a strong current passes through it. As a result, a hot plasma is formed in the course of propagation of the electric current, which provides transport of a large charge through this conductive channel and is accompanied by remarkable glowing. An individual flash lasts less than 10^{-3} s. During this time the conductive channel is heated and expands that creates a strong sound wave—thunder. Propagation of an electric current along the same channel may be realized several times.

Figure 5.4 contains the photography of typical lightning [118]. One can see the channel of lightning through which an electric current passes, and branches accompany this channel. If the leader of lightning propagates to the Earth and if outstanding objects are found at the surface and corona discharge are formed near them, then incoming currents move from these objects to the lightning leader (for example, [113, 119, 120]). As a result, the lightning leader is divided into several or many branches in the course of approaching to the Earth. In addition, several lightning leaders may be developed simultaneously.

As it is indicated above, lightning as electric breakdown of the atmosphere occurs under specific conditions. Namely, these conditions provide formation of dense cumulus clouds and the cloud temperature is large that leads to discharging of water microdroplets which form these clouds. Therefore, existence of lightning depends on the local geographical point. The maximal rates of lightning flashes are observed in the village of Kifuka of the Democratic Republic of Congo. This village is located

Fig. 5.4 Photography of some lightning [118]

at an altitude of 975 m above sea level, and the intensity of lightning flashes is 160 per km^2 and per year. In comparison, this value averaged over the globe is 3 per km^2 and per year.

One more example of the region with a higher probability of thunderstorms is the mouth of the Catatumbo river, which flows into Lake Maracaibo in Venezuela [122]. The Catatumbo region is surrounded from three sides by ridges of Ands and Cordilleras of 3700 m in height above the sea level. This landscape creates a specific character of air fluxes. In addition, methane is emerged from surrounding marshes, and its presence in the atmosphere facilitates lightning generation. Lightnings arise often between clouds and reach 10 km in length. Such thunderstorms are called often as the Catatumbo phenomenon [121, 123, 124] and an example is represented in Figs. 5.5 and 5.6. Usually, thunderstorms in this region are observed at night and they last approximately 10 hours from 150 days per year.

The total number of lightning flashes is 1.2×10^6 per year. Assuming that each flash transfers a charge of 5 C, the average one over the globe [105], one can obtain that the contribution of the Catatumbo region to the lightning current is 0.2 A compared to the total electric current to the Earth surface of 1700A in average. Because of a relatively high frequency and density of lightning discharges in the Catatumbo atmosphere, a large number of flashes may be observed there simultaneously, as it is shown in Fig. 5.6.

It should be noted that weak atmospheric processes which influence the atmosphere as an electric circuit act simultaneously on thermal processes in the atmosphere. This may lead to correlation between parameters of the atmospheric electric circuit and its thermal state (or climate), and this correlation takes place as a result

Fig. 5.5 Typical lightning of Catatumbo [121]

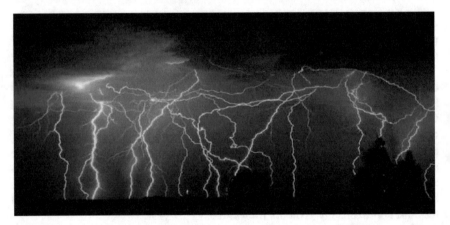

Fig. 5.6 Series of atmospheric breakdowns in Catatumbo [125]

of action of external factors which determine energetics of the atmosphere [48, 126–130]. From this, it follows also that local electric atmospheric parameters depend on certain conditions of a region under consideration. In particular, this can be expressed in the frequency of lightning act in this region. Indeed, the lightning rate for the Earth is approximately 44 ± 5 times per second, but it is distributed nonuniformly over the Earth's surface. Approximately 70% of lightning observations take place in a tropical part of land in accordance with the character of this process which proceeds in a warm season.

In considering physics of the lightning process, we restrict ourselves by three stages, namely, the first stage is formation of the conductive channel resulted from propagation of the ionization wave—a stepped leader, the second stage consists in the passage of the electric current through the conductive channel, and this stage establishes the equilibrium for the distribution of cloud charges with the ground, and the third stage is decay of the conductive channel. In reality, the second stage of this model, the recurrent stroke, corresponds to current from two sides. The electric current from the ground starts from prominent objects and then the conductive channel may be divided into branches. In particular, Fig. 5.4 contains the photography of a lightning discharge, and some branches for propagation of the lightning signal to the ground are observed there.

Along with the above stages of lightning development, there are intermediate stages and additional details of this phenomenon. The contribution of each stage to the lightning electric current may be different depending on the lightning intensity and conditions of its development. Our goal is to analyze the processes in the matter which is created as a result of this phenomenon and properties of this matter on the basis of observation data and physics of appropriate processes. Therefore, we are guided by lightning with average parameters. We use that the cloud electric potential with respect to the Earth in the thunderstorm weather ranges $U = (20 - 100)$ MV [103]. In subsequent estimations, we take $U = 60$ MV. For an altitude of clouds over the Earth $h = 3$ km, this gives the atmospheric electric field strength $E = U/h = 200$ V/cm that corresponds to its typical values in the thunderstorm weather.

A substance of the conductive channel is dissociated and partially ionized air. The temperature of the conductive channel is determined on the basis of spectroscopy measurements starting from 1868 [131]. A typical temperature of the conductive channel is (20,000–30,000) K or (2–3) eV. If intensities of some spectral lines are responsible for the air temperature, the number densities of electrons and ions follow from shifts of spectral lines for excited atoms and molecules. In particular, the Holzmark effect [132], i.e. the shift of spectral lines of the hydrogen atom under the action of an electric field created by ions, allows one to determine the ion number density in the region of maximum air ionization, that is, $\sim 10^{17} \text{cm}^{-3}$. Of course, lightning parameters vary in the course of passage of the electric current through the conductive channel, and we will use the above typical values for estimations.

One can see that lightning as an electric breakdown in atmospheric air occurs at low electric field strengths which is two orders of magnitude below the breakdown electric field strength. Hence, one can consider lightning as an ionization wave. Such a wave determines a specific electric breakdown in atmospheric air which takes place in sparks [69–72, 75, 76, 90, 113, 116]. Below we use the understanding of this phenomenon represented in indicated books to describe some aspects of their physics. The basis of electric breakdown in dense gases is a streamer as a solitary ionization wave [133, 134]. We demonstrate some peculiarities of a streamer which follow from numerical evaluations of this phenomena (for example, [135–139]).

Figure 5.7 contains streamer parameters which result from numerical evaluations [139] where the streamer propagates in atmospheric air. In this case, ionization proceeds through excited nitrogen molecules which are formed in electron–molecule

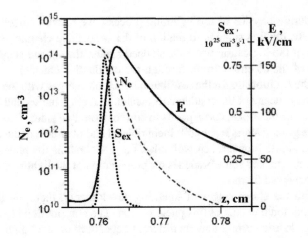

Fig. 5.7 Distribution of streamer parameters in air along the streamer axis before the ionization wavefront [139] : x is a distance from the streamer front, E is the electric field strength, N_e is the electron number density, S_{ex} is the rate of excitation of nitrogen molecules in the state $N_2(C^3\Pi_u)$ as a result of collisions between electrons and nitrogen molecules in the ground state

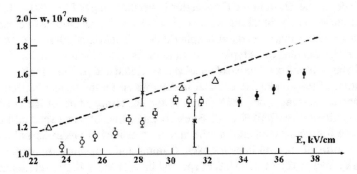

Fig. 5.8 Reduced to the atmospheric pressure experimental values of the electron drift velocity in air in the limit of a low electron number density near the breakdown electric field strength. Squares—[141], triangles—[142], crosses—[143], dotted line—[144]

collisions and then they transfer its excitation through resonant photon. This photon is absorbed at some distance from the source that leads subsequently to ionization there as a result of collisions with surrounded oxygen molecules. The cross section of photon absorption is approximately 10^{-18}cm^2 [140]. Hence, according to these experimental data, the mean free path of the resonant photon is 0.05 cm, i.e. it is absorbed on average at a distance ~0.02 cm from an emission point along the direction of streamer propagation because of isotopic character of emission.

According to Fig. 5.7 the region of formation and multiplication of electrons is narrow. In addition, Fig. 5.8 gives the electron drift velocity at the electric field strength in vicinity of the breakdown one. The real velocity of streamer propagation lies between the electron drift velocity and the light speed which characterizes the wave transfer through photoionization process. A streamer as an ionization wave propagates with the velocity $v_s \sim 10^8$ cm/s. This velocity corresponds to more complicated ionization waves, including lightning and the case of the streamer mechanism of the elementary ionization transfer. Then the current density for such an ionization wave is

$$i \sim ev_s N_e \sim 10\,\text{A/cm}^2$$

An intensive ionization wave, a leader, includes many streamers which act in the leader cross section simultaneously. The leader is responsible for ionization propagation in spark which physics is described in detail in books [69–72, 75, 76, 90, 113, 116]. We consider below some features the physics of the leader as a system of many streamers on the basis of [76, 116] with using additional information. In particular, we note that the leader as an ionization wave is surrounded by negative charges which support a certain radius of a moving leader.

Indeed, let us consider a chain of processes which involve electrons and accompany the leader propagation. When an electron goes under the region of the field action near the front of a propagating ionization wave, it is cooled as a result of vibration excitation of nitrogen molecules. A typical time τ_t of electron thermalization in this manner is estimated as

$$\tau_t \sim \frac{n}{N_m v_e \sigma_{ex}}$$

Here, $n \sim 10$ is a number of vibration excitations which lead to the electron thermalization, $N_m \sim 10^{19}$ cm^{-3} is the number density of nitrogen molecules, $v_e \sim 10^8$ cm/s is a thermal electron velocity, $\sigma_{ex} \sim 10^{-15}$ cm^2 [145, 146] is a typical cross section of vibration excitation of the nitrogen molecule in collisions with electrons at electron energies of a few electronvolts. From this it follows $\tau_t \sim 10$ps. Hence, a negatively charged case around the ionization wave is formed fast.

One more peculiarity of the leader propagation is a delay of establishment of the charge equilibrium behind the head of the ionization wave where the electric field strength is equal $E_c = 5$kV/cm according to measurements for sparks [116]. Electrons are located in this region, and after propagation the ionization wave these electrons return until the negative and positive charges behind the wave front would be equal. In accordance with Fig. 5.8, the drift velocity of electrons in the process of charge equalization behind the wavefront is $\sim 10^7$ cm/s, whereas the velocity of the wavefront is $\sim 10^8$ cm/s. As a result, the region length for a non-compensated negative charge grows with an increasing length of wave propagation. This effect takes place also in long sparks.

This mechanism acts in the case of the stepped leader as the first stage of the lightning process. Until electrons remain behind the ionization wave—leader, the electric field strength there is the same as at the beginning $E_c = 5$ kV/cm that is below the breakdown one. Hence, when the leader passes the path $U/E_c \sim 10$m, the electric field is closed. This is comparable with an observed length of an individual segment for the stepped leader. After passage of each segment, the ionization wave is stopped up to neutralization of the formed channel. After this, the electric voltage is restored, and the wave propagates along a new segment.

Let us pay an attention to the character of formation of the conductive lightning channel after passage of the stepped leader. Until air of the lightning channel is not heated, the ionization process of air molecules leads to formation the molecular ions, such as N_2^+ and O_2^+. The rate of dissociative recombination involving these ions and thermal electrons is approximately $\alpha = 2 \times 10^{-7}\,\mathrm{cm^3/s}$ (for example, [147]), and even values the electron number density are of the same order of magnitude as behind a streamer $N_e \sim 10^{14}\,\mathrm{cm^{-3}}$ (see Fig. 5.7), the recombination process is over during a small time $\tau_{\mathrm{rec}} \sim 0.1\,\mu\mathrm{s}$ compared to a time of passage of a distance between clouds and ground ($\sim$0.01 s). Hence, the lightning channel remains to be conductive if it is heated by a passed electric current. In addition, spectral analysis of the lightning flash according to which the electron temperature of the lightning channel is equal (2–3) eV.

In considering the charge transport from clouds to the Earth's surface and concentrating on key lightning processes, we are guided by a simplified lightning model which is present in Fig. 5.9. The electric current passed through the lightning conductive channel transfers a part of the cloud negative charge to the Earth's surface during a small time compared to the lifetime of this channel that is observed as a flash. After this the channel is decomposed, but until its conductivity is not small, subsequent current pulses may pass through this channel with intensive emission. Because charge transport is fast, when the current is locked, a specific configuration of charges is established at the ground and clouds, so that the negative charge uniformly located along clouds of sizes of several kilometers and for its shielding the positive charge is grouped near the throat of the conductive channel. In the same manner, a negative charge is gathered at the ground near the throat of the conductive channel.

After finishing the electric current through the conductive channel, two relaxation processes proceed, namely, the first one is the cooling of the conductive channel with a decrease of its conductivity, and the other process is a departure of an electric charge inside the ground. If the second relaxation process is slower than the first one, a repeated electric breakdown of air is possible. A typical time of relaxation of the conductive lightning channel is of the order of $t_o \sim 0.01$ s, and the relaxation time of spreading of the ground charge near the channel throat is determined by the ground conductivity. If the ground conductivity is enough high, several flashes

Fig. 5.9 Simple lightning model [102] : 1—charged cloud, 2—metal rod for transferring a charge from a charged cloud to the Earth, 3—Earth surface

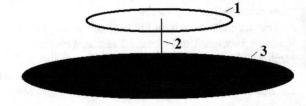

of lightning are possible, whereas at low ground conductivity repeated lightning flashes are impossible. Below we determine the boundary conductivity Σ_o of the ground which separates these regimes of lightning.

Within the framework of the scheme of Fig. 5.9, after gathering of the flash charge near the throat of the conductive channel, it is spreading over an underground hemisphere. Equation of departure of an underground charge Q as a result of a current I from the conductive throat is given by

$$I = -\frac{dQ}{dt} = 2\pi E(R)\Sigma \cdot 2\pi R^2, \tag{5.10}$$

where a current electric field strength $E(R)$ on a distance R from a basis of the conductive channel is

$$E = \frac{Q}{R^2}$$

we use the Ohm law for an underground current, and Q is the charge inside the hemisphere. Solution of equation (5.10) has the form

$$Q = Q_o \exp(-2\pi \Sigma t)$$

Here, Q_o is a transported charge through the lightning channel, or a charge at the base of the conductive channel at the beginning, i.e. at time $t = 0$.

From this, we have the criterion of the lightning regime with several flashes

$$\Sigma \gg \frac{1}{2\pi t_o} \tag{5.11}$$

Taking $t_o \sim 0.01$ s, we obtain according to this criterion $\Sigma \gg 10^{-13}$ S/m. This criterion is fulfilled for a ground of any composition and is not valid for the atmosphere. Hence, usually lightning has several flashes as a result of charge transport through a ground.

5.2.2 Properties of the Conductivity Lightning Channel

Propagation of a strong electric current through the conductive lightning channel located between a cloud and ground occupies the central place in lightning physics. The matter of this channel is created by lightning and exists during a small time providing pass of the electric current in a short time. Below we consider physical properties of this matter that is dissociated and partially ionized air. Because of high pressure, components of this ionized gas are found in the local thermodynamic equilibrium. Then according to the dissociative equilibrium, one half of nitrogen molecules in atmospheric air is dissociated at the temperature 6000 K, and for oxygen molecules this temperature is approximately 4000 K. Considering temperatures above 6000 K, we have the ionization equilibrium in this plasma in the form

$$e + N^+(^3P) \leftrightarrow N(^4S), \; e + O^+(^4S) \leftrightarrow O(^3P) \tag{5.12}$$

On the basis of the Saha distribution [148, 149] for the equilibrium (5.12), we have the following relation between the number density of nitrogen atoms $[N]$, oxygen atoms $[O]$, nitrogen atomic ions $[N^+]$, and oxygen ions $[O^+]$

$$\frac{N_e[N^+]}{[N]} = \frac{g_e g(N^+)}{g(N)} \left(\frac{m_e T}{2\pi\hbar^2}\right)^{3/2} \exp\left(-\frac{J_N}{T}\right), \; \frac{N_e[O^+]}{[O]} = \frac{g_e g(O^+)}{g(O)} \left(\frac{m_e T}{2\pi\hbar^2}\right)^{3/2} \exp\left(-\frac{J_O}{T}\right) \tag{5.13}$$

Here, the statistical weights are $g(N) = 4$, $g(O) = 9$, $g(N^+) = 9$, $g(O^+) = 4$ for indicated atoms and ions, and $J_N = 14.534 \text{eV}$, $J_O = 13.618 \text{eV}$ are the ionization potentials for these atoms. It is convenient to reduce (5.13) to one equation

$$\frac{N_e^2}{N_a} = \left[\frac{9}{2}c_N \exp\left(-\frac{J_N}{T}\right) + \frac{8}{9}c_O \exp\left(-\frac{J_O}{T}\right)\right] \left(\frac{m_e T}{2\pi\hbar^2}\right)^{3/2}, \tag{5.14}$$

where $N_a = [N] + [O]$ is the total number density of atoms in dissociated air, c_N and c_O are the concentrations of nitrogen and oxygen atoms in dissociated atmospheric air; for simplicity, we neglect a contribution of argon to this equilibrium. From this, one can represent the electron concentration in hot atmospheric air as

$$c_e = \frac{N_e}{N_a + 2N_e} \tag{5.15}$$

Figure 5.10 gives the temperature dependence for the electron concentration in hot equilibrium air in accordance with formulas (5.14) and (5.15). One can see that total ionization of equilibrium atmospheric air proceeds at temperatures above 20.000 K. In addition, we consider the time range above a time of establishment of the ionization

Fig. 5.10 Electron concentration c_e defined by formula (3.6) for the ionization equilibrium in dissociated atmospheric air

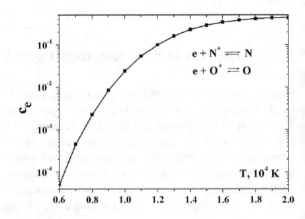

equilibrium grows with a decreasing temperature and is $\sim 0.1s$ at $T = 5000\,\text{K}$ at atmospheric pressure. Below we consider temperatures of atmospheric air above $5000\,\text{K}$.

We now determine the conductivity Σ of air located in the lightning conductive channel on the basis of a general formula [150]

$$\Sigma = \frac{N_e e^2 \tau}{m_e}, \tag{5.16}$$

where N_e is the electron number density, m_e is the electron mass, τ is a typical scattering time for electrons. We account for electron-atom and electron-ion scattering mechanisms for electrons, and contributions of these scattering channels to the total conductivity are comparable at the electron concentration $\sim 1\%$ [151]. A typical scattering time is estimated as $1/\tau = N_a v_T \sigma_{ea} + N_i v_T \sigma_{ei}$, where N_a is the number density of air molecules, N_i is the number density of ions, v_T is a typical thermal electron velocity, σ_{ea} is the diffusion cross section of electron-atom scattering, and σ_{ei} is the cross section of electron-ion scattering. Assuming scattering parameters to be identical for oxygen and nitrogen atoms, one can obtain for the conductivity of this air

$$\Sigma = \left(\frac{1}{\Sigma_{ea}} + \frac{1}{\Sigma_{ei}} \right)^{-1}, \tag{5.17}$$

where Σ_{ea} is the conductivity due to electron-atom scattering, and Σ_{ei} is the conductivity due to electron-ion scattering. At low temperatures the conductivity of this air is determined by electron-atom scattering, while at high temperatures the main contribution to the air conductivity follows from electron-ion scattering.

Formula (5.16) gives for the air conductivity due to electron-atom scattering

$$\Sigma_{ea} = \frac{N_e e^2 \tau_{ea}}{m_e}, \quad \frac{1}{\tau_{ea}} = N_a v_e \sigma_{ea}, \tag{5.18}$$

where N_a is the number density of atoms, v_e is a typical electron velocity, σ_{ea} is the diffusion cross section of electron–atom collisions. Being guided by electron temperatures $T_e \sim 10,000$ K, we have for a typical electron velocity $v_e \sim 1 \times 10^7 \text{cm/s}$ and a typical cross section of electron-atom scattering is equal [152] $\sigma_{ea} \approx 2 \times 10^{-16} \text{cm}^2$. From this we have the following estimation for the conductivity of dissociated atmospheric air due to electron-atom scattering

$$\Sigma_{ea} = \frac{N_e \Sigma_o}{N_a \sqrt{T_e}}, \tag{5.19}$$

where $\Sigma_o = 2 \times 10^6 \text{S/m}$, and the electron temperature in formula (5.19) is expressed in eV.

At high temperatures, the conductivity of dissociated air is determined by electron-ion scattering and is given by formula [76, 153]

$$\Sigma_{ei} = \frac{\Sigma_1 T_e^{3/2}}{\ln \Lambda} \tag{5.20}$$

Here, $\Sigma_1 = 1.9 \times 10^4$ S/m, the electron temperature is expressed in eV, and the Coulomb logarithm $\ln \Lambda$ is taken as $\ln \Lambda = 7$. Then measuring the conductivity in S/m, one can represent formula (5.20) in the form

$$\Sigma_{ei} \approx 3 \times 10^3 T_e^{3/2} \tag{5.21}$$

Combining formulas (5.19) and (5.21) on the basis of relation (5.17), one can give the conductivity of atmospheric dissociated air as a temperature function in the temperature range under consideration that is present in Fig. 5.11. The transition from the mechanism of electron-atom scattering to the mechanism of electron-ion scattering for atmospheric air proceeds at the temperature $T = 7300$ K, and the conductivity of atmospheric air is $\Sigma = 3 \times 10^3$ S/m at this temperature.

Because of exchange by heat between the conductive channel and air around it, the channel gets cool and is broaden after current pass. In the range of air temperatures and channel radii under consideration, the heat exchange is realized through air thermal conductivity at atmospheric pressure. Therefore, we determine below the thermal conductivity coefficient κ of dissociated air that characterizes heat transport across the lightning channel. Heat transport under these conditions is determined by two mechanisms, namely, by collisions between air atoms, as well as by electron-atom collisions. Correspondingly, the total thermal conductivity coefficient κ is a sum of the thermal conductivity coefficients due to transport of atoms κ_a and electrons κ_e, i.e.

$$\kappa = \kappa_a + \kappa_e \tag{5.22}$$

Fig. 5.11 Conductivity of dissociated air at atmospheric pressure in the temperature range typical for the lightning channel is determined by electron-atom and electron-ion scattering in accordance with formula (5.17)

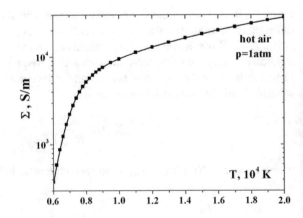

In determination the thermal conductivity coefficient κ_a due to atom transport, we use the model of hard spheres (for example, [154]). Within the framework of this model, collision of two atoms is analogous to collisions of hard balls, and the partial conductivity coefficient in this case is given by (for example, [155, 156])

$$\kappa_a = \frac{75\sqrt{\pi T}}{64\sigma_g\sqrt{m}}, \tag{5.23}$$

where T is the temperature, m is the atom mass, σ_g is the gas-kinetic cross section for collision of two atoms. Dissociated air consists of nitrogen and oxygen atoms located in the ground states ($N(^4S)$ and $O(^3P)$). The statistical weights of these atom states are $g_N = 4$ and $g_O = 9$. A number of electron terms for interaction of each atom pair, i.e. a number of degenerate electron states of the quasi-molecule consisting of colliding atoms, is large. Therefore, a gas-kinetic cross section σ_g in formula (5.23) corresponds to averaging over many channels.

At high temperatures, this cross section is determined by the approach of colliding atoms at which their electron shells are overlapped. In this case, we use the analogy with collisions of two neon atoms, and then the gas-kinetic cross section for collisions of nitrogen or oxygen atoms is approximately $\sigma_g = 19\,\text{Å}^2$ at room temperature [155]. Correspondingly, if we take the gas-kinetic cross section for collisions of nitrogen or oxygen atoms as one half of the gas-kinetic cross sections for collision of nitrogen or oxygen molecules, one can obtain approximately the same value for the gas-kinetic cross section of atoms. In addition, we use below the temperature dependence for the gas-kinetic cross section has the form $\sigma_g \sim T^{-0.2}$ as it follows from an average over various states [155].

In the analysis of heat transport in air, we operate with the thermal diffusivity coefficient χ that is an analog of the diffusion coefficient for particle transport and is equal to

$$\chi = \frac{\kappa}{c_p N_a} \tag{5.24}$$

where $c_p = 5/2$ is the heat capacity per atom at constant pressure, N_a is the atom number density. Figure 5.12 contains the thermal diffusivity coefficient of dissociated air owing to atom transport.

The thermal conductivity coefficient due to electron transfer in an equilibrium plasma with identical gaseous and electron temperatures ($T_e = T$) is given by [154, 157]

$$\kappa_e = \frac{2c_e\sqrt{T}}{3\sqrt{\pi m_e}\sigma_{ea}} = \frac{0.532c_e\sqrt{T}}{\sigma_{ea}\sqrt{m_e}}, \tag{5.25}$$

where c_e is the electron concentration, σ_{ea} is the diffusion cross section of electron-atom scattering which assumes to be independent of the collision velocity. We take this cross section to be $\sigma_{ea} = 3 \times 10^{-16}\,\text{cm}^2$ on the basis of data for electron scattering in various gases [152].

Formula (5.25) corresponds to low electron and ion concentrations where one can neglect electron-ion scattering. In order to take into account the contribution of electron-ion scattering to the thermal conductivity coefficient, we use the analogy in electron-ion scattering for the conductivity coefficient of this plasma and for thermal conductivity coefficient. Then the contribution to the thermal conductivity coefficient due to electron-ion collision one can determine by the factor $\Sigma_{ea}/(\Sigma_{ea} + \Sigma_{ei})$, where Σ_{ea}, Σ_{ei} are the air conductivities which are given by formulas (5.19) and (5.21) correspondingly. As a result, we have for the thermal diffusivity coefficient of dissociated and partially ionized air due to electron transfer as

$$\chi_e = \frac{0.532\Sigma_{ea}c_e\sqrt{T}}{(\Sigma_{ea} + \Sigma_{ei})\sigma_{ea}\sqrt{m_e}}, \tag{5.26}$$

Figure 5.12 contains the thermal diffusivity coefficient of dissociated air χ_e due to electron transfer. As is seen, though a thermal velocity of electrons exceeds that for atoms significantly, the cross section of electron-ion scattering with Coulomb interaction is larger than that for electron-atom and atom-atom scattering. For this reason, the contribution of electron transfer to the thermal diffusivity coefficient is small in a wide range of temperatures.

On the basis of the above parameters of atmospheric dissociated air, we estimate those ones for the lightning channel. Taking its conductivity $\Sigma \sim 2 \times 10^4$ S/m, as it follows from Fig. 5.11, and accounting for a typical electric field strength $E = 200$ V/cm for the thunderstorm weather, one can estimate a typical current density passed through conductivity channel

$$i = \Sigma E \sim 5 \times 10^4 \, \text{A/cm}^2$$

Let us use the relation which connects a typical pulse duration τ, a typical charge Q passed through the channel, and the channel radius ρ

Fig. 5.12 Thermal diffusivity coefficient in dissociated air due to electron scattering and atom scattering

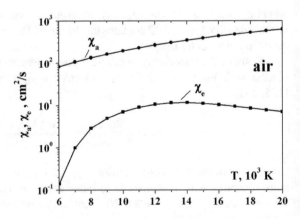

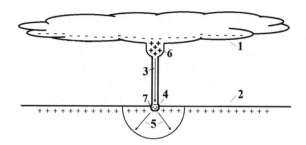

Fig. 5.13 Character of propagation of an electric current of lightning underground. 1—cloud, 2—ground, 3—conductive lightning channel, 4—charge transferred underground, 5—propagation of current underground, 6—cloud throat of the conductive channel, 7—ground ending of the conductive channel

$$Q = \pi \rho^2 i \tau$$

Taking a typical charge of one flash of middle lightning [105] $Q = 5\,C$, one can obtain the following relation for parameters of one flash of middle lightning

$$\pi \rho^2 \tau \sim 10^{-4}\,\text{cm}^2\,\text{s} \tag{5.27}$$

In the course of propagation of a stepped leader as an ionization wave, the following connection takes place between a time τ of leader propagation and a current radius square ρ^2 for the conductive channel

$$\rho^2 = 4\chi\tau, \tag{5.28}$$

where we assume that in the course of channel expansion its temperature $T \sim$ 20.000 K conserves, and $\chi(20.000\,\text{K}) = 600\,\text{cm}^2/\text{s}$. One can add to this that a typical time of leader propagation from a cloud to the Earth is $\tau_p = L/v_l$, where a typical speed of leader propagation is $v_l \sim 10^7\,\text{cm/s}$, and a typical distance between a cloud and Earth is $L \sim 3\,\text{km}$, i.e. a typical time of leader propagation is $t_p \sim 0.03\,\text{s}$. Hence, the radius of the lightning conductive channel according to formula (5.28) $\rho > 8\,\text{cm}$. Returning to formula (5.26), one can conclude from this that the duration of the current pulse for middle lightning is measured in μs.

Let us consider other aspects of lightning development. According to observations, lightning is not dominate channel for charge transport from cumulus clouds to ground. In addition, lightning development starts when the charge of clouds transfers from microdroplets to moleculae ions. But even in this case the cloud charge does not partake in propagation of the lightning current. Indeed, if these ions are molecular ones with the mobility approximately $2\,\text{cm}^2/(\text{V} \cdot \text{s})$, their drift velocity in the atmospheric field of strength $E = 200\,\text{V/cm}$ is $4\,\text{m/s}$, i.e. during existence of the lightning conductive channel they displace on small distances. Hence, lightning develops under the electric field of clouds and ground only and results in transport of a positive charge to a throat of the conductive channel. Subsequently, this charge is transported to other cloud parts relatively long.

In order to analyze one more aspect of lightning as a charge transport, we represent in Fig. 5.13 an electric scheme of lightning. Because the charge transported through the conductivity channel is accumulated at its endings in the course of current pass, this charge creates an electric field strength in the opposite direction with respect to the atmospheric one. In the end, this field will lock the atmospheric one. Hence, a transported charge is restricted by the value q_{max} that is given by

$$q_{max} < \frac{Eh^2}{4} \tag{5.29}$$

Taking the cloud-ground distance (the altitude of clouds) $h = 3\,km$ and the atmospheric electric field strength E, one can obtain the maximum charge $q_{max} = 7\,C$ which can be collected at endings of the conductive channel.

5.2.3   Heat Regime of the Conductivity Lightning Channel

We now consider the energetics in formation an ionized and dissociated air being guided by the temperature of the conductive channel of $T = 20.000$ K. At this temperature, each air molecule is dissociated and almost ionized. Taking for simplicity that all atoms are ionized (see Fig. 5.12), one can find the energy ε_c consumed per one atomic particle (electron or ion)

$$\varepsilon_c = \frac{D}{2} + c_p T + \frac{J}{2},$$

where D is the molecule dissociation energy, J is the atom ionization potential, $c_p = 5/2$ is the heat capacity per one electron or ion at the constant pressure, and the final temperature $T = 20.000$ K is large compared to the initial one. From this, we obtain roughly $\varepsilon_c = 20\,eV$, and the energy consumed per unit volume for ionization inside air at atmospheric pressure and room temperature is approximately $1\,J/cm^3$. One can compare it with the specific energy EQ which is extracted after passage of the electric current, so that $E = 200\,V/cm$ is the electric field strength, and Q is the transferred charge. Taking $Q = 5\,C$, one can obtain the energy per unit length of the conductive channel $1\,kJ/cm$. As is seen, parameters of atmospheric air and lightning allow one to create the conductivity lightning channel.

Being guided by the relaxation process, where after passage of the electric pulse a plasma of the conductive channel of lightning is cooled, we are based on the model with adiabatic cooling of a hot plasma of the conductive channel. For simplicity, we ignore the heat extracted as a result of recombination process and assume that the channel has a sharp boundary, as well as a temperature variation inside it. Then the heat of a hot region is proportional to $\rho^2 T$, where ρ is a current radius of the channel.

Under these assumptions we have $\rho^2 T = \rho_o^2 T_o$, where ρ_o is the initial radius of the conductive channel, T_o is the initial temperature. This allows one to rewrite equation (5.27) in the form

$$\frac{d\rho^2}{dt} = 4\chi(T)$$

or

$$\frac{d(1/T)}{dt} = \frac{4\chi(T)}{\rho_o^2 T_o}$$

After propagation of the electric current almost all air in the channel is ionized. Subsequently, when the current stops, electrons and ions recombine, and until their concentration is remarkable, this channel may be used for the current one more. According to lightning measurements, this time is approximately 0.2 s. We now analyze the kinetics of this recombination and also of temperature relaxation. The three body recombination process involving electrons and ions proceeds according to the scheme

$$2e + A^+ \rightarrow e + A^* \tag{5.30}$$

The recombination coefficient of this process may be determined on the basis of the Thomson theory [158] according to which a third particle carries an excess energy, and colliding particles, an electron and ion in this case, forms a bound state. In this case, the three body recombination coefficient K is given by [159]

$$K_{\text{rec}} = C \frac{e^{10}}{m_e^{1/2} T_e^{9/2}}, \tag{5.31}$$

where e is the electron charge, m_e is the electron mass, T_e is the electron temperature, C is a numerical coefficient, and this formula requires fulfillment of the criterion

$$N_e e^6 / T^3 \ll 1$$

The numerical coefficient C is equal [147, 160] $C = 2 \times 10^{\pm 0.3}$. At the temperature $T_o = 20.000$ K the recombination coefficient is equal $K(T_o) = 4.5 \times 10^{-34} \text{cm}^6/\text{s}$. According to the definition, for this recombination process the three body recombination coefficient K_{rec} in a quasi-neutral plasma is included in the balance equation

$$\frac{dN_e}{dt} = -K_{\text{rec}} N_e^3 \tag{5.32}$$

We assume above the adiabatic character of expansion of the conductive channel being guided by the range of parameters where atmospheric air of the conductivity channel is dissociated and partially ionized. Because we start from the temperature

$T_o = 20.000$ K, where air is almost entirely ionized, in the course of a temperature decrease the heat release takes place in a not wide temperature range. In particular, at the temperature $T_o = 20.000$ K, the three body coefficient of recombination of electrons and ions is $K_{rec} = 4.5 \times 10^{-28}$ cm^6/s, and the number density of electrons and ions is $N_e = 1.7 \times 10^{17}$ cm^{-3} that corresponds to a typical recombination tine $\tau_{rec} \sim 10^{-8}$ s. Since we deal with a time range $(10^{-4} - 10^{-3})$ s, one can accept a prompt recombination of electrons and ions. Hence, in the above formula the temperature T_o must be changed by T_1 which is given by formula

$$T_1 = T_o + \frac{J}{2c_p}$$

Thus, we consider above the adiabatic character of expansion of the conductive channel, starting from the temperature $T_1 \approx 5$ eV, and the above equation for a temperature decrease has the form

$$\frac{dT}{dt} = \frac{4\chi(T)T^2}{\rho_o^2 T_1} \tag{5.33}$$

Combining (5.32) and (5.33), one can obtain the equation which connects the number density N_e of electrons or ions and the temperature T of the conductive channel

$$\frac{dN_e}{N_e^3} = -d\left(\frac{1}{T}\right)\frac{\rho_o^2 T_o K_{rec}(T)}{4\chi(T)} \tag{5.34}$$

Accounting for the temperature dependence $\chi(T) \sim T^{0.7}$ and $K_{rec}(T) \sim T^{-4.5}$, one can obtain the connection between the electron number density N_e and the temperature T of the conductive channel

$$N_e(T) = \frac{N_o}{\sqrt{1 - B\left[1 - \left(\frac{T}{T_o}\right)^{6.2}\right]}}, \quad B = \frac{N_o^2 \rho_o^2 K_{rec}(T_o)}{12\chi(T_o)}, \tag{5.35}$$

where N_o is the initial number density of electrons, i.e. $N_o = N(T_o)$, and taking $T_o = 20.000$ K, we have $N_o = 1.8 \times 10^{17}$cm^{-3}, $K_{rec}(T_o) = 4.5 \times 10^{-28}$cm^6/s, $\chi(T_o) = 600$cm^2/s, $B/\rho_o^2 = 2.0 \times 10^{-3}$.

In the same manner, one can express the parameters of plasma cooling through the reduced time

$$\tau = \frac{t}{t_o}, \quad t_o = \frac{\rho_o^2}{6.8\chi(T_o)}, \tag{5.36}$$

where $t_o(T_o)/\rho_o^2 = 2.4 \times 10^{-4}$ s/cm^2. Plasma parameters are equal in the course of its cooling

$$N_e = \frac{N_o}{\sqrt{1 - B(\tau^{3.65} - 1)}}, \quad T = \frac{T_o}{\tau^{0.6}}, \quad \rho = \rho_o \tau^{0.3} \tag{5.37}$$

In addition, on the basis of formula (5.19) with using these data one can determine the conductivity of the lightning channel in the course of its evolution. Figure 5.13 contains the results of these calculations for relaxation of the lightning conductive channel after passage of the electric current pulse.

Note that in this analysis we assume violation of the dissociative equilibrium in the course of thermal relaxation of the lightning channel. Let us check this assumption basing on the rate constant of the three body process

$$2N + N_2 \rightarrow 2N_2 \tag{5.38}$$

is found in the range $(1–2) \times 10^{-34} \, cm^6/s$ at temperatures $(6000–8000)$ K [161, 162], and the number density of atoms in this temperature range is $(3–5) \times 10^{17} \, cm^{-3}$ in this temperature range. From this, one can find a typical time of three body recombination of atoms that leads to molecule formation $\tau_r \sim (0.01 - 0.1)$ s and is compatible with a typical time of thermal relaxation.

Let us analyze these results. We use the simple model where a plasma of the conductive lightning channel and surrounding atmospheric air are separated by a sharp boundary, and expansion of this channel in the course of its relaxation after passage of the current pulse results from heat transport through the thermal conductivity of the plasma. In addition, expansion of the conductive channel proceeds adiabatically, i.e. this plasma does not obtain an additional energy during its expansion. We also take into account a heat release due to recombination of electrons and ions, but formation of molecules as a result of association of nitrogen and oxygen atoms is ignored because of larger times of association compared to times under consideration. Therefore, in spite of the qualitative character of this analysis, the model under consideration allows to understand the behavior of the relaxing plasma of the conductive channel.

We take as a time scale of Fig. 5.14 small times compared with those ones between neighboring flashes for middle and strong lightning. During times $\lesssim 0.1$s the degree of air ionization is supported which provides passing of a subsequent electric pulse. As it follows from these Figures, the first stage with recombination of the most part of electrons and ions proceeds fast for times ~ 0.01s and a subsequent evolution of the conductive channel consists in its slow expansion and relaxation of weakly ionized nonequilibrium air. This relaxation proceeds slower for wider channels, and it is demonstrated in Table 5.2 where parameters of the plasma of the conductive channel are given for 0.1 s after passage of the electric current through a given cross section. From this, it follows also that in the case of several flashes, a subsequent flash proceeds under more favorable conditions than those for the previous one. Moreover, as it follows from Table 5.2, the relative variation of the radius of the conductive channel decreases with an increase of the initial radius.

It should be noted that thermal relaxation of the lightning channel leads simultaneously to the ionization relaxation of this channel. Indeed, at low temperatures

Fig. 5.14 Parameters of relaxation of the lightning conductive channel after passage of the electric current at the initial temperature $T = 20.000$ K for equilibrium ionized air in this channel. 1—the initial radius of the conductive channel is $\rho_o = 1$ cm, 2—the channel radius is $\rho_o = 2$ cm at the beginning. These parameters evaluated according to formula (5.37) are the electron (and ion) number density (**a**), the temperature of atoms, electrons and ions (**b**), a current radius of the conductive channel (**c**). The conductivity of air in the conductive channel (**d**) is calculated by formula (5.19)

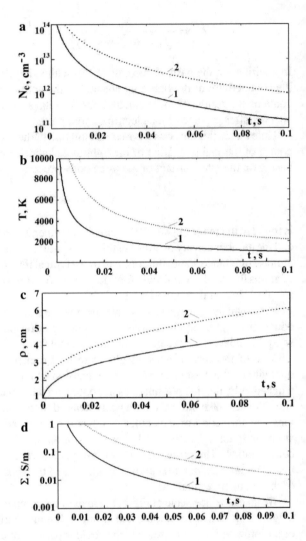

molecular ions are formed, and their recombination with electrons proceeds fast because of the dissociative recombination is a pair process. The boundary temperature is (4000–5000) K, and below this temperature recombination involving electrons and ions is accelerated. Therefore, the lightning channel with a high conductivity is conserved until the air temperature is not low. Thus, the nature chooses an optimal way of charge transport between clouds and ground through creation a conductive channel with a high temperature. This channel provides propagation through it several subsequent flashes if a time between flashes is less than a time of thermal relaxation of dissociated air in this conductivity channel.

Let us return to Fig. 5.13 which contains an electric scheme of lightning. As it follows from this figure, transport of charge is caused by a large electric potential of

Table 5.2 Parameters of relaxing air of the lightning conductive channel within the framework of the model of a sharp boundary between the conductive channel and surrounding air through 0.1 s after passage of an electric current. The initial temperature of equilibrium ionized air of the conductive channel is 20000 K. Here, ρ_o is the initial (after the electric pulse) channel radius, N_e is the number density of electrons (and ions), T is the air temperature (identical for electrons, ions, and atoms), ρ is the final radius of the conductive channel, Σ is the final air conductivity due to electrons

ρ_o, cm	1	2	3	5
N_e, 10^{12} cm^{-3}	0.17	1.0	2.8	9.4
T, 10^3 K	0.94	2.1	3.4	6.0
ρ, cm	4.6	6.1	7.2	9.1
Σ, S/m	1.5×10^{-3}	0.015	0.049	0.21

a cloud with respect to a ground. When the propagation of the lightning current is over, a positive charge is concentrated near the cloud throat of the conductive channel and the same negative charge is collected at the ground ending of the channel. After this dissociated air is cooled and its conductivity decreases. During this time ions of the cloud shift on a distance of the order of tens meters that is small compared to the distance between a ground and cloud. But the charge of the ground ending can disappear if the ground conductivity is high. In this case, a new flash is possible.

References

1. E. Williams, Atmosp. Res. **91**, 140 (2009)
2. C.T.R. Wilson, Proc. Roy. Soc. **92A**, 555 (1916)
3. C.T.R. Wilson, Philos. Trans. Roy. Soc. **221A**, 73 (1921)
4. C.T.R. Wilson, Proc. Roy. Soc. **37A**, 32D (1925)
5. H. Israël, *Atmospheric Electricity* (Keter Press Binding, Jerusalem, 1973)
6. C.B. Moore, B. Vonnegut, in *Lightning*, ed. R.H. Golde (Academic Press, London, 1977). p. 51
7. M. Rycroft et al., Space Sci. Rev. **137**, 83 (2008)
8. M.J. Rycroft et al., in *Planetary Atmospheric Electricity*, ed. by F. Leblanc, et al. (Springer, Heidelberg, 2008), p. 83
9. R.P. Feynman, R.B. Leighton, M. Sands, *The Feynman Lectures of Physics*, vol. 2 (Addison-Wesley, Reading, 1964)
10. B.M. Smirnov, *Cluster Processes in Gases and Plasmas* (Wiley, Berlin, 2010)
11. J.A. Chalmers, *Atmospheric Electricity* (Claredon Press, Oxford, 1949)
12. Y.I. Frenkel, *Theory of Phenomenon of Atmospheric Electricity* (GITTL, Leningrad, 1949). in Russian
13. B.F.J. Schonland, *Atmospheric Electricity* (Methuen, London, 1953)
14. B.F.J. Schonland, *The Lightning Discharge. Handbuch der Physik*, vol. 22 (Springer, 1956), p. 576
15. H. Israël, *Atmospheric Electricity*, vol. 1 (Conductivity (Academische Verlagsgesellschaft, Leipzig, Fundamentals, 1957)
16. H. Israël, *Atmospheric Electricity*, vol. 2 (Charges, Currents (Academische Verlagsgesellschaft, Leipzig, Fields, 1961)

17. J.A. Chalmers, *Atmospheric Electricity* (Pergamon Press, Oxford, 1967)
18. R.G. Harrison, Surv. Geophys. **25**, 441 (2004)
19. C. Haldoupis, M. Rycroft, E. Williams, C. Price, J. Atmos. Solar-Terr. Phys. **164**, 127 (2017)
20. J.H. Kraakevik, The airborne measurement of atmospheric conductivity. J. Geophys. Res. **63**, 161 (1958)
21. J.F. Clark, Airborne measurement of atmospheric potential gradient. J. Geophys. Res. **62**, 617 (1957)
22. C.T.R. Wilson, J. Franklin Inst. **208**, 1 (1929)
23. G.C. Simpson, Mon. Weather Rev. **34**, 16 (1906)
24. W. Lowrie, *Fundamentals of Geophysics* (Cambridge University Press, Cambridge, 2007)
25. https://en.wikipedia.org/wiki/Global-atmospheric-electrical-circuit
26. S. Israelsson, Pure and Appl. Geophys. **116**, 149 (1978)
27. D. Singh, V. Gopalakrishnan, R.P. Singh et al., Atmosp. Res. **84**, 91 (2007)
28. R.G. Harrison, Surv. Geophys. **34**, 209 (2013)
29. S.J. Mauchly, Note on the diurnal variation of the atmospheric electric potential gradient. Phys. Rev. **18**, 161 (1921)
30. S.J. Mauchly, (1923) On the diurnal variation of the potential gradient of atmospheric electricity. Terr. Magn. **28**, 61 (1923)
31. B.F.J. Schonland, The polarity of thunderclouds. Proc. Roy. Soc. **118A**, 233 (1928)
32. F.J.W. Whipple, Quart. J. Roy. Met. Soc. **55**, 351 (1929)
33. T.W. Wormell, Atmospheric electricity: some recent trends and problems. Quart. J. Roy. Met. Soc. **79**, 474–489 (1953)
34. F. Märcz, R.G. Harrison, Ann. Geophys. **23**, 1987 (2005)
35. E.R. Williams, Lightning and climate: a review. Atmos. Res. **76**, 272 (2005)
36. E.R. Williams, R. Markson, S. Heckman, Geophys. Res. Lett. **32**, L19810 (2005)
37. F. Märcz, R.G. Harrison, Geophys. Res. Lett. **33**, L12803 (2006)
38. R. Reiter, *Phenomena in Atmospheric and Environmental Electricity* (Elsevier, New York, 1992)
39. http://en.wikipedia.org/wiki/Siemens
40. M. Uyeshima, Surv. Geophys. **28**, 199 (2007)
41. N. Olson, A. Kuvshinov, Earth Planet Sci. **56**, 525 (2004)
42. S. Israelsson, E. Knudsen, Effect of radioactive fallout from a nuclear power plant accident on electrical parameters. J. Geophys. Res. **91**, 11909 (1986)
43. S. Israelsson, T. Schutte, E. Pisler, S. Lundquist, Increased occurrence of lightning flashes in Sweden during 1986. J. Geophys. Res. **92**, 10996 (1987)
44. A. Simon, Nucl. Sci. Abstr. **66**, 146 (1962)
45. C.D. Stow, Atmospheric electricity. Rep. Prog. Phys. **32**, 1–67 (1969)
46. R. Markson, J. Geophys. Res. **90 D**, 5967(1985)
47. E.J. Adlerman, E.R. Williams, J. Geophys. Res. **101 D**, 29679(1996)
48. M.J. Rycroft, S. Israelsson, C. Price, J. Atmos. Sol. Terr. Phys. **62**, 1563 (2000)
49. B.A. Tinsley, G.B. Burns, L. Zhou, The role of the global electric circuit in solar and internal forcing of clouds and climate Adv. Space Res. **40**, 1126–39 (2007)
50. B.A. Tinsley, The role of the global electric circuit in solar and internal forcing of clouds and climate. Rep. Prog. Phys. **71**, 066801 (2008)
51. W.O. Schumann, Zeit. Naturforschung **A7**, 149 (1952)
52. W.O. Schumann, Zeit. Naturforschung **A7**, 250 (1952)
53. https://en.wikipedia.org/wiki/Schumann-resonances
54. *Handbook of Atmospheric Electrodynamics* ed. by H. Volland (CRC Press, Boca Raton, 1995)
55. A.P. Nickolaenko, M. Hayakawa, Y. Hobara, Long-term periodical variations in the global lightning activity deduced from the Schumann resonance monitoring. J. Geophys. Res. **104**, 27585 (1999)
56. A. Nickolaenko, M. Hayakawa, *Resonances in the Earth-ionosphere Cavity* (Kluwer, Dordrecht, 2002)
57. E.R. Williams, Science **256**, 1184 (1992)

58. https://en.wikipedia.org/wiki/Francis-Hauksbee
59. M.B. Schiffere, *Draw the Lightning Down: Benjamin Franklin and Electrical Technology in the Age of Enlightenment* (California University Press, Berkeley, 2003)
60. F. Ronalds, *Catalogue of Books and Papers Relating to Electricity, Magnetism, the Electric Telegraph, Etc.* ed. by A.J. Frost. (Cambridge, Cambr.Univ.Press, 2013)
61. C.F.C. Dufay. Philos. Trans. **38**, 258 (1734)
62. *Dictionary of the History of Science* ed. by W.F. Bynum, E.J. Browne, R. Porter (Princeton University Press, Princeton, New Jersey, 1981)
63. W.M. Saslow, *Electricity, Magnetism, and Light* (Texas, Thomson, 2010)
64. *Experiments and Observations on Electricity Made at Philadelphia in America. By Benjamin Franklin* (Good Press, 2019)
65. https://en.wikipedia.org/wiki/Michael-Faraday
66. https://en.wikipedia.org/wiki/Julius-Pluecker
67. https://en.wikipedia.org/wiki/Johann-Wilhelm-Hittorf
68. https://en.wikipedia.org/wiki/J.J.Thomson
69. L.B. Loeb, *Fundamental Processes of Electrical Discharges in Gases* (Wiley, New York, 1939)
70. J.M. Meek, J.D. Craggs, *Electrical Breakdown of Gases* (Claredon Press, Oxford, 1953)
71. H. Raether, *Electron Avalanches and Breakdown in Gases* (Butterworth, London, 1964)
72. F. Llewellyn-Jones, *Ionization and Breakdown in Gases* (Methuen, London, 1966)
73. M. Mitchner, C.H. Kruger, *Partially Ionized Gases* (Wiley, New York, 1973)
74. J.A. Rees, *Electrical Breakdown in Gases* (Macmillan, London, 1973)
75. L.B. Loeb, J.M. Meek, *Mechanism of Electric Spark* (Stanford University Press, Stanford, 1941)
76. E.M. Bazelyan, Y.P. Raizer, *Spark Discharge* (CRC Press, Roca Baton, 1998)
77. https://en.wikipedia.org/wiki/Electric-spark
78. E. Flegler, H. Raether, Zs. Tech. Phys. **16**, 435 (1935)
79. H. Raether, Phys. Zs. **15**, 560 (1936)
80. H. Raether, Zs. Tech. Phys. **18**, 564 (1937)
81. H. Raether, Zs. Phys. **107**, 91 (1937)
82. H. Raether, Zs. Phys. **110**, 611 (1938)
83. H. Raether, Zs. Phys. **112**, 464 (1939)
84. H. Raether, Zs. Phys. **117**, 375 (1941)
85. H. Raether, Ergebn. Ex. Naturwissenshaft **22**, 73 (1949)
86. L.B. Loeb, J.M. Meek, J. Appl. Phys. **11**, 438 (1940)
87. L.B. Loeb, J.M. Meek, J. Appl. Phys. **11**, 459 (1940)
88. L.B. Loeb, Phys. Rev. **81**, 287 (1951)
89. L.B. Loeb, Phys. Rev. **94**, 227 (1954)
90. J.M. Meek, J.D. Crags, *Electrical Breakdown in Gases* (Wiley, New York, 1998)
91. J.M. Meek, Proc. Phys. Soc. **52**, 547 (1940)
92. J.M. Meek, Proc. Phys. Soc. **52**, 822 (1940)
93. J.M. Meek, Phys. Rev. **57**, 722 (1940)
94. J.S. Townsend, Electrifician **50**, 971 (1903)
95. J.J. Thomson, *Conduction of Electricity Through Gases* (Cambridge University Press, Cambridge, 1904)
96. J.S. Townsend, *Electricity in Gases* (Claredon Press, Oxford, 1915)
97. J.S. Townsend, *Motion of Electrons in Gases* (Claredon Press, Oxford, 1925)
98. H. Tholl, Zs. Phys. **172**, 536 (1963)
99. H. Tholl, Zs. Naturforsch. **19a**, 346 (1964)
100. https://en.wikipedia.org/wiki/Kerr-cell-shutter
101. B.M. Smirnov, Phys. Usp. **57**, 1041 (2014)
102. B.M. Smirnov, *Microphysics of Atmospheric Phenomena* (Springer Atmospheric Series, Switzerland, 2017)

103. K. Berger, The Earth flash, in *Lightning*, ed. R.H. Golde. (Academic Press, London, 1977), p. 119
104. D. Malan, *Physics of Lightning* (The English University Press Ltd., London, 1963)
105. M.A. Uman, *Lightning* (McGrow Hill, New York, 1969)
106. V.A. Rakov, M.A. Uman, *Lightning, Physics and Effects* (Cambridge University Press, Cambridge, 2003)
107. R. Clausius, Annalen der Physik **155**, 500 (1850)
108. W. Kenneth, *Thermodynamics* (McGraw-Hill, New York, 1988)
109. M.A. Uman, *About Lightning* (Dover, New York, 1986)
110. M.A. Uman, *The Lightning Discharge* (Academic Press, New York, 1987)
111. V.A. Rakov, *Fundamental of Lightning* (Cambridge University Press, Cambridge, 2016)
112. E.V. Cooray, *The Lightning Flash* (Institution of Engineering and Technology, London, 2003)
113. Yu.P.Raizer, *Physics of Gas Discharge.* (Dolgoprudnyi, Intellect, 2009; in Russian)
114. V. Cooray, *An Introduction to Lightning* (Springer, Dordrecht, 2015)
115. E.M. Bazelyan, Y.P. Raizer, Phys. Uspekhi **43**, 753 (2000)
116. E.M. Bazelyan, Y.P. Raizer, *Lightning Physics and Lightning Protection* (IOP Publishing, Bristol, 2000)
117. J.R. Dwyer, M. Uman, Phys. Rep. **534**, 147 (2014)
118. https://cdn.fishki.net/upload/post/2017/01/06/2184403/are-noteworthy-34-129.jpg
119. N.L. Alexandrov, E.M. Bazelyan, Y.P. Raizer, Phys. Plasma **31**, 84 (2005)
120. https://en.wikipedia.org/wiki/Lightning
121. http://maxcls.ya.ru/replies.xml?item-no=369
122. http://worldis.org/molnii-katatumbo-mayak-marakaybo-venesuela
123. http://en.wikipedia.org/wiki/Catatumbo-lightning
124. http://www.guardian.co.uk/world/2010/mar/05/venezuela-lightning-el-nino
125. http://www.amusingplanet.com/2010/06/mysterious-venezuela-catatumbo.html
126. R.G. Harrison, K.S. Carslaw, Rev. Geophys. **41**, 1012 (2003)
127. C. Price, Geophys. Res Lett. **20**, 1363 (1993)
128. E.R. Williams, Mon. Weather Rev. **122**, 1917 (1994)
129. E.R. Williams, Atmos. Res. **76**, 272–287 (2005)
130. A.J. Bennett, R.G. Harrison, Weather **62**, 277–283 (2007)
131. R.A. Herschel, Proc. Roy. Soc. **15**, 61 (1868)
132. J. Holtsmark, Ann. der Physik. **58**, 577 (1919)
133. https://en.wikipedia.org/wiki/Streamer-discharge
134. X. Lu, G.V. Naidis, M. Laroussi, K. Ostrikov, Phys. Rep. **540**, 123 (2014)
135. N.Y. Babaeva, G.V. Naidis, J. Phys. **29D**, 2423 (1996)
136. A.A. Kulikovsky, J. Phys. **30D**, 441 (1997)
137. A.A. Kulikovsky, Phys. Rev. **57E**, 7066 (1998)
138. A.V. Shelobolin, J. Phys. **40D**, 6669 (2007)
139. G.V. Naidis, Phys. Rev. **79E**, 057401 (2009)
140. N. Liu, V.P. Pasko, J. Geophys. Res. **109**, A04301 (2004)
141. K.R. Allen, K. Philips, Proc. Roy. Soc. **274A**, 163 (1963)
142. J.S. Townsend, H.T. Tizard, Proc. Roy. Soc. **A88**, 336 (1913)
143. W. Riemann, Zs. Phys. **122**, 216 (1944)
144. R.A. Nielsen, N.E. Bradbury, Phys. Rev. **51**, 69 (1937)
145. H.S.W. Massey, *Negative Ions* (Cambridge University Press, Cambridge, 1976)
146. B.M. Smirnov, *Negative Ions* (McGrow Hill, New York, 1982)
147. B.M. Smirnov, *Physics of Ionized Gases* (Wiley, New York, 2001)
148. M.N. Saha. Proc. Roy. Soc. **99 A**, 135(1921)
149. L.D. Landau, E.M. Lifshitz, *Statistical Physics*, vol. 1 (Pergamon Press, Oxford, 1980)
150. C. Kittel, *Introduction to Solid State Physics* (Wiley, New York, 1986)
151. B.M. Smirnov, *Plasma Processes and Plasma Kinetics* (Wiley, Berlin, 2007)
152. J. Dutton, J. Chem. Phys. Ref. Data **4**, 577 (1975)
153. L. Spitzer, *Physics of Fully Ionized Gases* (Wiley, New York, 1962)

154. B.M. Smirnov, *Principles of Statistical Physics* (Wiley VCH, Berlin, 2006)
155. B.M. Smirnov, *Reference Data on Atomic Physics and Atomic Processes* (Springer, Heidelberg, 2008)
156. M. Capitelli, D. Bruno, A. Laricchiuta, *Fundamental Aspects of Plasma Chemical Physics* (Springer, Transport (New York, 2013)
157. B.M. Smirnov, *Physics of Ionized Gases* (Mir, Moscow, 1981)
158. J. Thomson, Philos. Mag. **47**, 334 (1924)
159. L.M. Biberman, V.S. Vorob'ev, I.T. Iakubov, *Kinetics of Nonequilibrium Low Temperature Plasma* (New York, Consultants Bureau, 1987) §6
160. B.M. Smirnov, *Nanoclusters and Microparticles in Gases and Vapors* (DeGruyter, Berlin, 2012)
161. S. Byron, J. Chem. Phys. **44**, 1378 (1966)
162. J.P. Appleton, M. Steinberg, D.J. Liquornik, **48**, 599 (1968)

Chapter 6
Greenhouse Phenomenon in the Earth's Atmosphere

Abstract The "line-by-line" method is used for evaluation of thermal emission of the standard atmosphere toward the Earth. Accounting for thermodynamic equilibrium of the radiation field with air molecules and considering the atmosphere as a weakly nonuniform layer, we reduce emission at a given frequency for this layer that contained molecules of various types to that of a uniform layer which is characterized by a certain radiative temperature T_ω, an optical thickness u_ω and an opaque factor $g(u_\omega)$. Radiative parameters of molecules are taken from the HITRAN database, and an altitude of cloud location is taken from the requirement of coincidence of the total radiative flux from such evaluation with that followed from the energetic balance of the Earth. As a result of this evaluation for the contemporary atmosphere, we find that the radiative flux due to H_2O molecules equals $165 \, W/m^2$, the flux of $94 \, W/m^2$ is created by clouds, the radiative flux due to CO_2 molecules is $61 \, W/m^2$, CH_4 molecules create a flux of $4 \, W/m^2$, and the flux $4 \, W/m^2$ is due to N_2O molecules. In addition, approximately 95% of the radiative flux at frequencies below $800 \, cm^{-1}$ is created by H_2O and CO_2 molecules, while 84% of this flux at frequencies above $800 \, cm^{-1}$ is due to water microdroplets of clouds. It is shown that an increase of the concentration of one component which leads to an increasing radiative flux due to this component causes simultaneously to decreasing radiative flux due to other components because of overlapping of their spectra that corresponds to the Kirchhoff law. In particular, doubling of the concentration of atmospheric carbon dioxide gives an increase of the radiative flux due to this component by $7.2 \, W/m^2$, whereas radiative fluxes due to water molecules and water microdroplets decrease by $3.0 \, W/m^2$ and $2.9 \, W/m^2$ correspondingly, i.e. the change of the total radiative flux is $1.3 \, W/m^2$. This fact is not taken into account in some climatological models. Interaction of infrared radiation with water microdroplet and its pass through clouds is analyzed on the basis of the Mie model according to which a droplet is characterized by a sharp boundary. It is shown that stratus clouds, rather than cumulus ones, partake mostly in greenhouse phenomena of the Earth's atmosphere.

© The Editor(s) (if applicable) and The Author(s), under exclusive license
to Springer Nature Switzerland AG 2020
B. M. Smirnov, *Global Atmospheric Phenomena Involving Water*,
Springer Atmospheric Sciences, https://doi.org/10.1007/978-3-030-58039-1_6

153

6.1 Character of Atmospheric Emission Toward the Earth

6.1.1 Emission of Gaseous Layer

We first formulate the algorithm to evaluate the radiative flux toward the Earth from the atmosphere in the infrared spectrum range on the basis of general principles of emission from atmospheric air and transport of radiation in it [1–12]. One can represent the radiating atmosphere as a weakly nonuniform gaseous layer located over the Earth surface. Indeed, the atmospheric thickness for infrared radiation (~ 10 km) is small compared to the Earth radius ($R = 6370$ km), and a variation of the tropospheric temperature is relatively small. In this analysis, we take into account a high density of atmospheric air. Therefore, the radiation field is found in local thermodynamic equilibrium with air, and we use thermodynamic laws [13–16] in this analysis.

The Kirchhoff law [17] or the principle of detailed balance between the processes of emission and absorption is of importance also for this consideration. In particular, this principle allows us to analyze the emission process on the basis of the absorption coefficient k_ω that is the parameter of the absorption process. The absorption coefficient k_ω is introduced on the basis of the Beer–Lambert law [18, 19] according to which the energy flux I_ω of photons of a given frequency that propagates in a direction z is described as

$$\frac{dI_\omega}{dz} = -k_\omega I_\omega \qquad (6.1)$$

Here the radiation is noncoherent, i.e. formula (6.1) corresponds to an average over photon phases. In addition, the radiative flux is weak and does not influence the molecule behavior in the gas. Another parameter of the gaseous layer is the optical thickness u_ω of a layer of a thickness h is given by

$$u_\omega = \int_0^h k_\omega(h) dh, \qquad (6.2)$$

where h is the layer thickness.

Note that they are based on the "line-by-line" model [1] in this consideration. This implies that radiation is noncoherent that allows us to sum up the radiative intensities I_ω from individual radiators, as well as to represent the total radiative intensity as the sum of intensities at each frequency. As a result, one can represent the absorption coefficient k_ω as a sum of those for molecules and particles of the atmosphere. In addition, for pressures under consideration, the local thermodynamic equilibrium takes place in the atmosphere [15, 16, 20–23], and thermodynamic equilibrium takes place between atmospheric molecules and particles, as well as between the radiation field and atmospheric molecules.

Let us consider firstly emission of an uniform atmospheric layer where the temperature T is identical over the layer. Then the yield radiative flux I_ω at a given frequency ω is given by the Planck formula which describes the emission from the surface of a blackbody and has the form [15, 21]

$$I_\omega(T) = \frac{\hbar\omega^3}{4\pi^2 c^2 \left[\exp\left(\frac{\hbar\omega}{T}\right) - 1\right]}, \quad u_\omega \gg 1 \tag{6.3}$$

This means that a uniform gaseous layer emits as a blackbody at a large optical thickness. The frequency dependence for the radiative flux from a blackbody of an indicated temperature is represented in Fig. 6.1 for temperatures which are of interest for problems under consideration.

One can replace the frequency ω by the wavelength $\lambda = 2\pi c/\omega$ and the intensity in a new variable I_λ such that

$$I_\lambda d\lambda = I_\omega d\omega$$

We then have

$$I_\lambda = \frac{4\pi^2 c^2 \hbar}{\lambda^5 \left[\exp\left(\frac{2\pi\hbar c}{\lambda T}\right) - 1\right]} \tag{6.4}$$

The maximum of the value I_λ according to the Wien law [24] is realized at the wavelength

$$\lambda_{\max} T = 0.29 \, \text{cm} \cdot \text{K} \tag{6.5}$$

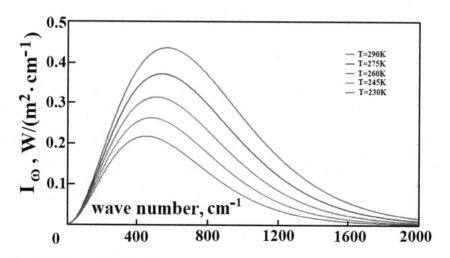

Fig. 6.1 Character of radiative flux from a blackbody

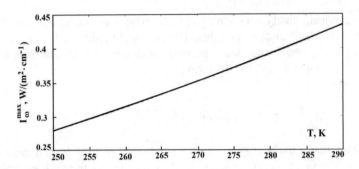

Fig. 6.2 Temperature dependence for the maximum radiative flux from a blackbody

Note that the value I_ω given by formula (6.3) takes place at another wavelength

$$\frac{\hbar\omega_{max}}{T} = 2.82 \tag{6.6}$$

If the equilibrium intensity of radiation (6.3) is realized at each frequency, the total radiative flux J is given by the Stefan–Boltzmann law

$$\int I_\omega d\omega = \sigma T^4, \tag{6.7}$$

where $\sigma = 5.67 \times 10^{-8}$ W/(m$^2 \cdot$ K^4) is the Stefan–Boltzmann constant. In addition, the maximum value of the partial radiative flux I_ω under equilibrium conditions according to formula (6.3) and formula (6.6) may be represented in the form

$$I_\omega^{max} = \gamma\sigma T^3, \ \gamma = 1.79 \times 10^{-8} \frac{\text{W}}{\text{m}^2 \cdot \text{K}^3 \cdot \text{cm}^{-1}} \tag{6.8}$$

Figure 6.2 contains the temperature dependence of I_ω^{max} for an equilibrium radiation and temperatures which are of interest for atmospheric emission.

We now determine the radiative flux at a given frequency J_ω for any optical thickness of a weakly nonuniform gaseous layer. In this case, it is necessary to sum intensities from elementary acts of emission as it is given in Fig. 6.3. Let us account for the isotropic character of an elementary radiative act and introduce an element of the solid angle as $d\Omega = d\cos\theta d\varphi$, where θ and φ are the polar and azimuthal angles. Summing the intensities from each point which intersects the boundary at the same point, one can obtain the yield radiative flux J_ω

$$J_\omega = 2I_\omega \int_0^1 d\cos\theta \int_0^{u_\omega} du_h \exp\left(-\frac{u_h}{\cos\theta}\right) = 2I_\omega \int_0^1 \cos\theta d\cos\theta \left[1 - \exp\left(-\frac{u_\omega}{\cos\theta}\right)\right] \tag{6.9}$$

Fig. 6.3 Geometry of photon propagation from a given plane of a thickness dh. The radiative flux from this plane is the sum of elementary fluxes

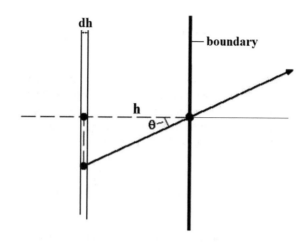

where u_ω is the total optical thickness of the layer (6.2), and u_h is a current optical thickness, i.e.

$$u_\omega = \int_0^L k_\omega(h)dh, \quad u = \int_0^h k_\omega(h')dh',$$

where L is the total thickness of the atmosphere. As a result, the radiative flux at a given frequency for a uniform gaseous layer is given by [16, 25]

$$J_\omega = I_\omega g(u_\omega), \quad g(u_\omega) = 2\int_0^1 \cos\theta d\cos\theta\left[1 - \exp\left(-\frac{u_\omega}{\cos\theta}\right)\right] \qquad (6.10)$$

The opaque factor $g(u_\omega)$ for a uniform gaseous layer is equal [16, 25]

$$g(u_\omega) = [1 - 2E_3(u_\omega)], \quad E_3(x) = \int_1^\infty \exp(-tx)\frac{dt}{t^3} \qquad (6.11)$$

In reality, the opaque factor is the average probability for a photon of a given frequency to reach the layer boundary if it is emitted isotropically from another boundary and is directed to this one.

In considering the emission of a weakly nonuniform gaseous layer, we assume the temperature to be dependent on a distance h from the boundary only. Emission from each gaseous point is characterized by the temperature of this point, and then using the Planck formula (6.3) for the radiative flux, one can represent it in the form [26, 27]

$$J_\omega = \frac{\hbar\omega^3}{2\pi^2 c^2} \int\limits_0^1 \cos\theta \, d(\cos\theta) \int\limits_0^{u_\omega} du \exp\left(-\frac{u}{\cos\theta}\right) F(u), \quad F(u) = \left\{\exp\left[\frac{\hbar\omega}{T(h)}\right] - 1\right\}^{-1},$$

$$(6.12)$$

and the function $F(u)$ depends on the temperature weakly.

We now consider an elementary volume according to Fig. 6.1 which is between two planes parallel to the boundary and located at a distance dh each from other. The optical thickness due to this layer is equal to $du_\omega = k_\omega dh$. Then the expression for the radiative flux J_ω which is formed inside an indicated plane is given by

$$J_\omega = 2I_\omega[T(h)]du_\omega \int\limits_0^1 \cos\theta \, d\cos\theta \exp\left[-\frac{u(h)}{\cos\theta}\right] \qquad (6.13)$$

We assume above that the temperature depends only on a distance h from the boundary.

Accounting for a weak temperature dependence $T(h)$ on a distance h from the boundary, one can expand this expression over a small parameter. We fulfill first this operation for an optically dense gas $u_\omega \gg 1$ and represent this expansion in the form

$$F(u_\omega) = F(u_o) + (u_\omega - u_o)F'(u_o) + \frac{1}{2}(u_\omega - u_o)^2 F''(u_o)$$

Taking the parameter u_o such that the second term of the expansion vanishes after integration that gives [26, 27]

$$J_\omega = I_\omega(T_\omega)(1 - \alpha); \quad u_o = u(h_\omega) = 2/3; \quad T_\omega = T(h_\omega); \quad \alpha = \left[\frac{5F''(u_o)}{18F(u_o)}\right]$$

$$(6.14)$$

where I_ω is given by formula (6.3).

In the above operation, we reduce emission of a weakly nonuniform gaseous layer to that of an uniform one. As a result, an effective point is introduced which is located at a distance h_ω from the boundary and determines emission of the total layer. In the case of an optically thick layer, the position of the effective point is given by $u(h_\omega) = 2/3$ and the radiative temperature coincides with that of a given point. In the same manner, in the other limiting case of an optically thin layer $u_\omega \ll 1$ the effective point for layer emission is given by [28, 29]

$$u(h_\omega) = u_\omega/2$$

From this in a general case, we have in the first approximation [28, 29]

$$J_\omega = I_\omega(T_\omega)g(u_\omega), \quad T_\omega = T(h_\omega), \quad u(h_\omega) = \frac{u_\omega}{2\exp(-u_\omega) + 1.5u_\omega} \qquad (6.15)$$

Thus, parameters through which is expressed the emission of a weakly nonuniform gaseous layer are the optical thickness u_ω of this layer, the opaque factor $g(u_\omega)$ of this layer, and the effective radiative temperature. These parameters allow one to determine the radiative flux at a given frequency on the basis of formula (6.15), and the total radiative flux results from integration of this quantity over frequencies.

Expression (6.15) for the yield radiative flux holds true for a monotonic distribution of radiators in a space. In a real atmosphere, these radiators are molecules of water and carbon dioxide, as well as microdroplets of water which form clouds. The number densities of optically active molecules in a space decrease monotonically at removal from the Earth's surface, whereas clouds are absent at low altitudes. Basing on this, we use the model where clouds are located starting from certain altitudes, and the optical thickness of clouds is large. Taking the atmosphere temperature T_{cl} at an altitude where clouds are located, one can generalize the expression (6.15) for the radiative flux to the form

$$J_\omega = I_\omega(T_\omega)g(u_\omega) + I_\omega(T_{cl})[1 - g(u_\omega)], \quad T_\omega = T(h_\omega), \quad u(h_\omega) = \frac{u_\omega}{2\exp(-u_\omega) + 1.5u_\omega}, \tag{6.16}$$

and the cloud temperature T_{cl} is independent of the frequency.

Formula (6.16) determines the radiative flux at a given frequency which is emitted by various layers of the atmosphere and is absorbed by the Earth's surface. We now formulate the transport equation in the atmosphere which accounts for emission in a certain atmospheric layer, as well as absorption in this layer for radiation created by other layers. Let us extract an atmosphere layer of a thickness dh or of an optical thickness $du_\omega = k_\omega dh$. The radiative flux dJ_ω at a frequency ω emitted by this layer is equal according to formula (6.12)

$$dJ_\omega^e = \frac{\hbar\omega^3 du_\omega}{2\pi^2 c^2} \int_0^1 \cos\theta d(\cos\theta) F(u_o) = 2I_\omega(T_o)du_\omega, \quad F(u_o) = \left[\exp\left(\frac{\hbar\omega}{T_o}\right) - 1\right]^{-1}, \tag{6.17}$$

where T_o is the temperature of this layer, $I_\omega(T)$ is the equilibrium radiative flux in accordance with formula (6.3). In the same manner, we determine the radiative fluxes emitted by the atmospheric molecules from layers below and above the layer under consideration on the basis of formula (6.15)

$$J_\omega^a = I_\omega(T_\omega^\uparrow)g(u_\omega^\uparrow)du_\omega + I_\omega(T_\omega^\downarrow)g(u_\omega^\downarrow)du_\omega, \tag{6.18}$$

where $T_\omega^\uparrow$ is the radiative temperature of the atmosphere part which is located above the layer under consideration, $T_\omega^\downarrow$ is that for the atmosphere part below this layer, $u_\omega^\uparrow$ and $u_\omega^\downarrow$ are the optical thicknesses of the atmosphere for its parts above and below this layer. It is necessary to add to this the radiative fluxes from the Earth's surface and from clouds in accordance with the model (6.16), where clouds are located at a certain altitude of the temperature T_{cl}, and its optical thickness at these altitudes varies sharply with the altitude change.

Let us introduce $(C_p dT/dt)_{\text{rad}}$ as the term in the heat balance equation of the atmosphere due to emission and absorption of infrared radiation. Here, C_p is the thermal capacity of atmospheric air per unit volume, T is the temperature of air located in this layer. Summing up the above expressions, we obtain

$$C_p \left(\frac{dT}{dt} \right)_{\text{rad}} = I_\omega(T_\omega^\uparrow) g(u_\omega^\uparrow) + I_\omega(T_\omega^\downarrow) g(u_\omega^\downarrow) + I_\omega(T_E)[1 - g(u_\omega^\downarrow)]$$

$$+ I_\omega(T_{cl})[1 - g(u_\omega^\uparrow)] - 2 I_\omega(T_o)\} k_\omega \tag{6.19}$$

This expression holds true for a weakly nonuniform atmospheric layer, and hence the optical thickness assumes to be independent of the temperature.

6.1.2 Thermal Emission of Atmosphere

Let us make the following step in realization the model (6.16) applying this scheme to the standard atmosphere model. Within the framework of the model of standard atmosphere with average parameters, the space distribution of the number density of water and carbon dioxide molecules is given by formulas (2.5) and (2.3). But determination of such distribution for water microdroplets is problematic, because formation of clouds has a random character. One can overcome this problem on the basis of the energetic balance of the Earth and its atmosphere which is given in Fig. 2.15.

Figure 6.2 presents infrared radiative fluxes in the energetic balance involving the Earth and atmosphere. It is of importance that fluxes toward the Earth and outside are separated that testifies about a large optical thickness of the atmosphere in the infrared spectrum range. This allows one to analyze emission of the atmosphere as an optically thick system. From the standpoint of the radiative model (6.16), it is possible to use the radiation energy balance of Fig. 6.4 for determination of the cloud temperature. Indeed, further along with formula (6.16) in determination of the fluxes of atmospheric emission, we use the relation for the total radiative flux toward the Earth

$$J_\downarrow \equiv \int J_\omega d\omega = 327 \ \text{W/m}^2 \tag{6.20}$$

At the first stage of this program, we consider a simple model [30] of atmospheric emission with a frequency-independent absorption coefficient k_ω and a large total optical thickness of the atmosphere. We also assume that the absorption coefficient decreases with an altitude monotonically and, for definiteness, approximates the altitude dependence for the absorption coefficient as

$$k_\omega = A \exp \left(-\frac{h}{\lambda} \right), \quad u(\infty) = A\lambda \tag{6.21}$$

Fig. 6.4 Average energy fluxes measured in W/m^2 and realized in the form of infrared radiation for the Earth and its atmosphere

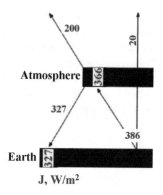

This model with data of Fig. 6.4 allows us to determine the radiative temperature which is independent now of the frequency and hence may be determined on the basis of the Stefan–Boltzmann law

$$J_{\downarrow} = \sigma T_{\downarrow}^4, \ J_{\uparrow} = \sigma T_{\uparrow}^4 \tag{6.22}$$

One can obtain for radiative temperatures of emission toward the Earth $T_{\downarrow}$ and outside $T_{\uparrow}$. We have now for these parameters [3, 30]

$$T_{\downarrow} = 276K, \ T_{\uparrow} = 244K \tag{6.23}$$

We also obtain for the altitudes which are responsible for atmospheric emission to the Earth $h_{\downarrow}$ and outside it $h_{\uparrow}$ on the basis of formula (2.2) that gives

$$T_{\downarrow} = T_E - \frac{dT}{dh} h_{\downarrow}, \ T_{\uparrow} = T_E - \frac{dT}{dh} h_{\uparrow},$$

where $T_E = 288$ K is the temperature of the Earth's surface for the model of standard atmosphere. From this, we find the altitudes which are responsible for atmospheric emission in this direction

$$h_{\downarrow} = 1.9 \, km, \ h_{\uparrow} = 6.8 \, km \tag{6.24}$$

These values allow us to determine the parameters of the approximation (6.21) on the basis of formula (6.14) according to which the atmospheric optical thickness from the effective layer to the boundary is 2/3. From this one can obtain [3, 30]

$$A = 0.46 \, km^{-1}, \ \lambda = 5.3 \, km \ u = 2.4; \tag{6.25}$$

Analyzing this model and its results, we note its roughness, because a monotonic dependence of the absorption coefficient on an altitude does not fulfill for clouds, and the frequency independence of the absorption coefficient is beyond criticism. Nevertheless, this allows us to estimate values of parameters of atmospheric emission. In particular, because inside the absorption bands of molecules, where absorption is strong, the radiative temperature is close to that of the Earth's surface, the cloud temperature is less than $T_\downarrow$ in formula (6.23), whereas the altitude h_{cl} for an atmospheric layer that is responsible for emission of clouds is larger than $h_\downarrow$ in formula (6.24). Roughly, from this one can take

$$T_{cl} < 270\,\text{K}, \quad h_{cl} > 3\,\text{km} \tag{6.26}$$

6.1.3   Molecular Emission of Atmosphere

Basing on the model (6.16) for atmospheric emission and transiting to radiation of the real atmosphere in infrared spectrum range, we take as atmospheric radiators three of its components, namely, molecules of water, and carbon dioxide, as well as water microdroplets which constitute clouds. In this model, one can separate in space molecules and clouds, so that radiating molecules are located in a gap between the Earth's surface and clouds. On the basis of the space distribution of CO_2 and H_2O molecules in accordance with formulas (2.3) and (2.5) with using parameters of molecule emission, one can determine the molecular part of the radiative flux in formula (6.16).

Note that along with H_2O and CO_2 molecules, one can account for other molecules, especially in specific cases of the atmosphere with a heightened concentration of some molecular components. But within the framework of the model of standard atmosphere in which average concentrations of molecular components are included, only atmospheric CH_4 and N_2O molecules create radiative fluxes to the Earth above $1\,\text{W/m}^2$. Therefore, we restrict only by these molecules considering them as impurities.

Thus, in the first approach, molecular components of the atmosphere are water vapor and carbon dioxide. In accordance with formulas (2.3), (2.5), and (2.6), the distribution of the number densities of carbon dioxide $N(CO_2)$ and water $N(H_2O)$ molecules over an altitude h is given by

$$N(H_2O) = N_w \exp\left(-\frac{h}{\lambda}\right), \quad N_w = 4.3 \times 10^{17}\,\text{cm}^{-3}, \quad \lambda = 2.00\,\text{km}$$

$$N(CO_2) = N_c \exp\left(-\frac{h}{\Lambda}\right), \quad N_c = 1 \times 10^{16}\,\text{cm}^{-3}, \quad \Lambda = 8.44\,\text{km} \tag{6.27}$$

In addition, the atmosphere temperature $T(h)$ as an altitude h function is

$$T(h) = T_E - \frac{dT}{dh}h, \quad T_E = 288\,\text{K}, \quad \frac{dT}{dh} = 6.5\,\text{K/km} \tag{6.28}$$

It should be noted that a nonuniformity of the atmosphere consists of different temperatures for different atmospheric layers. It leads to a frequency dependence for the radiative temperature in formula (6.16). As for the optical thickness u_ω of the atmosphere, in this consideration with using formula (6.27), we have for this value according to its definition (6.2)

$$u(h_\omega) = k_\omega(H_2O)\lambda\left[1 - \exp\left(-\frac{h_{cl}}{\lambda}\right)\right] + k_\omega(CO_2)\Lambda\left[1 - \exp\left(-\frac{h_{cl}}{\Lambda}\right)\right] \quad (6.29)$$

where h_{cl} is the altitude of cloud location, and $k_\omega(H_2O)$, $k_\omega(CO_2)$ are the absorption coefficients by water and carbon dioxide molecules near the Earth's surface.

In considering the absorption coefficient due to atmospheric molecules, accounting for the structure of the expression for this quantity [20, 22, 23, 33], it is convenient to represent it in the form [34]

$$k_\omega = N(H_2O)\sum_i S_i a(\omega - \omega_i) + N(CO_2)\sum_i S_i a(\omega - \omega_i), \quad (6.30)$$

where $N(H_2O)$ and $N(CO_2)$ are the total number densities of water and carbon dioxide molecules at a given point, S_i is the spectral line intensity for i-th transition between molecular states with the center at the frequency ω_i, $a(\omega - \omega_i)$ is the distribution function of emitted or absorbed photons over frequencies [20] which is normalized by the relation

$$\int a(\omega - \omega_i)d\omega = 1 \quad (6.31)$$

6.1.4 HITRAN Data Bank in Spectroscopy of Atmosphere

Our analysis of atmospheric emission is based on molecular spectroscopy and radiation thermodynamics. On the basis of laws and concepts of these fields of physics, one can connect the radiative fluxes from the atmosphere at each frequency with spectroscopic parameters of atmospheric components. Therefore, information about spectroscopic parameters of radiating molecules allows to transit from the academic description of atmospheric emission to the precise determination of radiative fluxes toward the Earth's surface.

Thus, we operate with intensities of individual transitions S_i which are proportional to the Einstein coefficients for these transitions, i.e. to their rates. Each intensity relates to a certain spectral line, and these values, as well as other parameters of spectral lines, are contained in the HIgh resolution TRANsmission (HITRAN) data bank [31, 32, 35]. This data bank contains information for radiative parameters of some molecules which are of interest for various problems (for example, [36, 37]). Information from this data bank is of importance in this algorithm for determination of

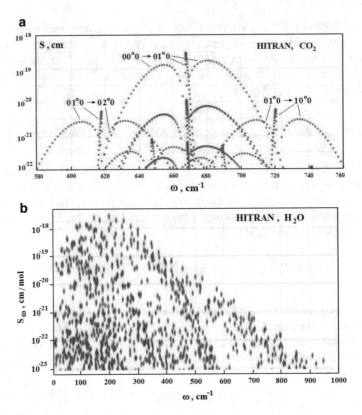

Fig. 6.5 Intensities for vibrational–rotational and rotational transitions which are responsible for thermal radiation of the atmosphere due to water molecules (**a**) and carbon dioxide molecules (**b**). The above data, taken from the HITRAN data bank [31, 32], refer to the temperature $T = 296\,\mathrm{K}$

radiative fluxes from the atmosphere. Figure 6.5 contains intensities of spectral lines for H_2O and CO_2 molecules [31, 32, 35].

The intensity of a spectral line due to a certain radiative transition of molecules is given by the expression [34]

$$S_i = \frac{\pi^2 c^2}{\omega_i^2} A_i \frac{g_i}{q(T)} \exp\left(-\frac{\varepsilon_i}{T}\right)\left[1 - \exp\left(-\frac{\omega_i}{T}\right)\right] \qquad (6.32)$$

Here, A_i is the first Einstein coefficient for this radiative transition, g_i is the statistical weight of the lower transition state, ε_i is the excitation energy of the lower transition state from the ground state of the molecule, T is the gas temperature, $q(T)$ is the statistical sum for this molecule. In particular, in the case of a linear molecule with a vibration-rotation radiative transition, we have

$$g_i = 2j + 1, \quad \varepsilon_i = \varepsilon_v + Bj(j+1), \quad q(T) = \frac{T}{B}\exp(-\varepsilon_v/T), \quad T \gg B \quad (6.33)$$

where j is the rotational quantum number for the lower state of the molecule, B is the rotational constant of the molecule, ε_v is the excitation energy for the ground rotational state and the given vibrational state v. Formula (6.32) may be used for CO_2 molecule.

Note that if the intensity $S_i(T_o)$ relates to the temperature T_o, one can represent the intensity $S_i(T)$ for a given vibration–rotation transition which is determined by

$$S_i(T) = S_i(T_o)\exp\left(\frac{\varepsilon_i}{T_o} - \frac{\varepsilon_i}{T}\right) \quad (6.34)$$

We take into account the strongest temperature dependence is the exponential one. In these evaluations for the atmosphere with the pressure of the order of atmospheric one, we use the impact mechanism of broadening of spectral lines, where the frequency distribution function of photons has the form

$$a_\omega = \frac{\nu_i}{2\pi[(\omega - \omega_i)^2 + (\nu_i/2)^2]}, \quad (6.35)$$

where the frequency ω_i refers to the center of the corresponding spectral line, and ν_i is its width. At atmospheric pressure adjacent lines overlap slightly, i.e. the following criterion holds true

$$\Delta\omega \gg \nu_i \quad (6.36)$$

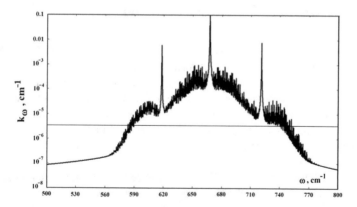

Fig. 6.6 Absorption coefficient in atmospheric air due to CO_2 molecules near the Earth's surface for the basic absorption band of carbon dioxide molecules at room temperature. The solid line corresponds to the atmosphere optical thickness $u_\omega = 2/3$ [29]

In particular, the absorption coefficient is given in Fig. 6.6 for atmospheric CO_2 molecules and the basic absorption band at room temperature and atmospheric pressure.

In this consideration, we are based on possibilities of contemporary molecular spectroscopy. Indeed, molecular spectroscopy starts together with development of quantum mechanics more than a century ago. The goal of this branch of quantum mechanics is to connect the structure of the molecule with parameters of rotation and vibration molecular transitions [38–47]. In this case, the theory extracts intensively the so-called dipole transitions, where the matrix element of the dipole moment operator between transition states of the molecule is nonzero. Then the rate of the radiative transition, as well as the absorption coefficient of a gas consisting of these molecules, are expressed through the Einstein coefficients for this transition, which are proportional to the square of an indicated matrix elements. This operation is transparent for molecules with the strong symmetry and is complicated in most cases.

In reality, in contemporary evaluations, they are based on data of banks, as the HITRAN database. This database includes information from theory and experiment, so that the selection rules and other connection between the molecule structure and parameters of radiative transitions are hidden inside the database. As a result, in practice, it is convenient to use information of the HITRAN database in certain evaluations (for example, [34, 48]). Nevertheless, the character of radiative transitions which is determined by the symmetry of transition states may be of interest for problems under consideration.

6.1.5 Spectrum of Atmospheric Emission

As it follows from Fig. 6.6, the basic absorption band due to CO_2 molecules is determined by vibration-rotation molecule transitions for three vibration transitions. Three resonances of Fig. 6.6 relate to the Q-branches of these vibration transitions, where the change of the rotation momentum of the molecule is zero. Because the criterion (6.33) holds true, the absorption spectrum has an oscillation character. The ratio of absorption coefficients for neighboring maximum and minimum is approximately 40 for the CO_2 molecule at atmospheric pressure, and roughly is 500 for the H_2O molecule.

In order to demonstrate the oscillation structure of molecular absorption spectra in the atmosphere, we give in Fig. 6.7 the absorption spectrum of the atmosphere near the Earth's surface due to CO_2 molecules in a narrow spectrum range. If we separate the infrared atmospheric spectrum due to H_2O and CO_2 molecules, one can represent the absorption coefficient of the atmosphere k_ω as

$$k_\omega = k_\omega(H_2O) + k_\omega(CO_2) \tag{6.37}$$

The next step in this program is determination of the radiative temperature T_ω due to molecules of the atmosphere on the basis of formula (6.16). In limiting case, we have for the radiative temperature

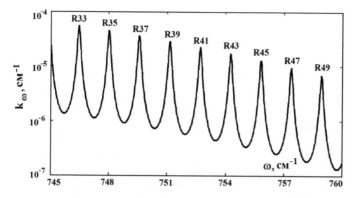

Fig. 6.7 Absorption coefficient in atmospheric air due to CO_2 molecules near the violet boundary of the absorption band. The absorption coefficient corresponds to the vibrational transition $01°0 \rightarrow 10°0$, and numbers indicate the initial rotational number of the CO_2 molecule for the rotation R—branch

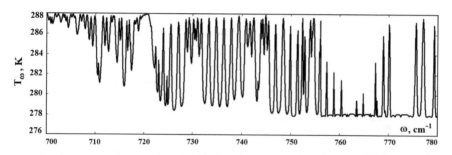

Fig. 6.8 Radiative temperature for emission toward the Earth created in the atmospheric gap between the Earth's surface and clouds due to CO_2 and H_2O molecules in the frequency range $(700–800)\,cm^{-1}$, if the gap size is $L = 3\,km$ [29]

$$T_\omega = T_E = 288\,K,\ u_\omega \gg 1;\ T_\omega = \frac{T_E + T_{cl}}{2},\ u_\omega \ll 1 \qquad (6.38)$$

As an example, Fig. 6.8 represents the radiative temperature T_ω in the frequency range $(700–780)\,cm^{-1}$ which includes the boundary of the absorption band for atmospheric carbon dioxide molecules. In principle, the radiative temperature depends on a distance L between the Earth's surface and clouds. Hence, it is necessary to summarize the partial radiative fluxes at different L and choose the optimal gap size which satisfies to the relation (6.20). But in reality, this operation is simplified because at large optical thicknesses due to molecules the radiative temperature is independent of L, and at frequencies, where the optical thicknesses of the atmosphere owing to atmospheric molecules the radiative flux toward the Earth is determined by clouds.

Let us divide the spectrum range into several parts. Figure 6.9 contains the total radiative fluxes to the Earth which are created by different components. As is seen, it is convenient to divide frequencies into two parts, below and above $800\,cm^{-1}$.

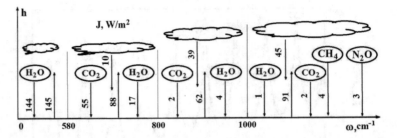

Fig. 6.9 Radiative fluxes in the infrared spectrum range from the Earth's surface in four frequency ranges, the fluxes toward the Earth due to clouds, and the fluxes from the atmosphere to the Earth due to indicated molecules inside some frequency ranges [29]. Corresponding values of radiative fluxes for indicated molecular components are given near arrows, average values of the fluxes are expressed in W/m^2

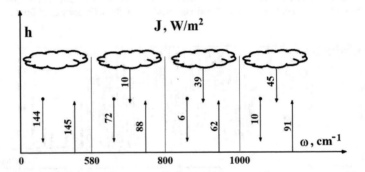

Fig. 6.10 Radiative fluxes in the infrared spectrum range from the Earth's surface, as well as those toward the Earth due to clouds and atmosphere inside indicated frequency ranges. Corresponding values of radiative fluxes are expressed in W/m^2

Roughly, at frequencies below $800\,cm^{-1}$ the radiative flux to the Earth is created by atmospheric molecules, whereas at larger frequencies clouds are the source of the radiative flux. This conclusion follows also from Fig. 6.9 in which data are obtained from those of Fig. 6.5. Next, Fig. 6.10 summarizes these fluxes and gives the total fluxes from the atmosphere due to their molecules, from clouds and the Earth.

Accounting for the above formulas, one can represent the absorption coefficient in the form

$$k_\omega = N \sum_i S_i(T_o) \frac{\nu_i}{2\pi[(\omega - \omega_i)^2 + (\nu_i/2)^2]} \exp\left(\frac{\varepsilon_i}{T_o} - \frac{\varepsilon_i}{T}\right), \qquad (6.39)$$

where N is the number density of H_2O or CO_2 molecules, and the summation is implied over all molecular components. The absorption coefficient is expressed here through four parameters for each spectral line, namely, ω_i is the transition frequency for the center of the spectral line, ν_i is the width of the spectral line, S_i is the intensity for the radiative transition under consideration, and ε_i is the excitation energy of

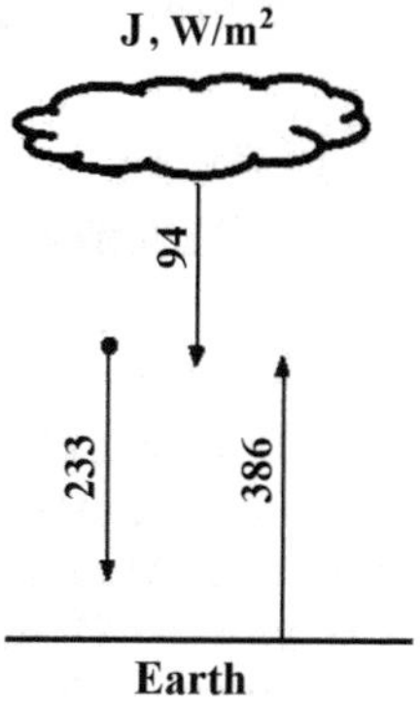

Fig. 6.11 Total radiative
fluxes in the infrared
spectrum range from the
Earth's surface, as well as
toward the Earth due to
clouds and atmosphere
expressed in W/m^2

Fig. 6.12 Specific radiative
flux per unit frequency from
the Earth's surface (1), and
its part reached clouds (2)
[29]

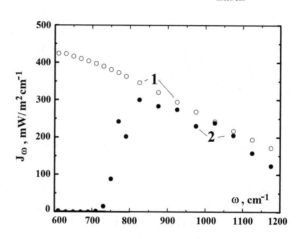

the lower state of the transition from the ground state of the molecule. All these parameters are contained in the tables "HITRAN on line" [31, 32].

One can see that determination of the absorption coefficient of the atmospheric air at a given frequency requires four parameters for some radiative transitions of atmospheric molecules which give contribution to the radiative flux according to formula (6.39). These parameters are contained in the HITRAN data bank, so that this data bank takes the key place in determination the radiative fluxes.

We note also the fact that atmospheric emission due to molecules dominates at low frequencies, whereas emission of the atmosphere at high frequencies is created by clouds. This is demonstrated in Fig. 6.12 where the radiative flux from the Earth's surface is compared with that reached clouds. From this Figure, it follows that the boundary frequency between these frequency ranges is close to $800 \, cm^{-1}$. An average is made in Fig. 6.12 over the frequency range of $20 \, cm^{-1}$ for frequencies below $800 \, cm^{-1}$, and this average is made over the frequency range of $50 \, cm^{-1}$ for frequencies above $800 \, cm^{-1}$.

6.2 Water in Atmospheric Radiation

6.2.1 Water Molecules and Microdroplets as Atmospheric Radiators

As it follows from the above analysis, three basic components are responsible for thermal emission of the atmosphere in the infrared spectrum range, namely, water and carbon dioxide molecules, as well as water microdrops. We now consider possibilities of water molecules as atmospheric radiators. In principle, they are more strong radiators than CO_2 molecules for two reasons. First, in contrast to the CO_2 molecule, the water molecule has the dipole moment, and hence its interaction with a radiation field is stronger. Second, the water concentration in a lower atmosphere is higher than that of carbon dioxide molecules. In particular, near the Earth's surface, the average concentration of water molecules is 1.7% compared to 0.04% of CO_2 molecules. In addition, free water molecules compose the basic water mass in the atmosphere, and the average humidity of the atmosphere is approximately 80% at the ground level. Therefore at temperatures which correspond to energies of vibration-rotation or rotation transitions water molecules have to determine emission of atmospheric air.

We now represent the spectroscopic parameters of a free H_2O-molecule [38, 49]. This molecule has the C_{2v}-symmetry, i.e. it consists of two identical segments OH in the ground electron state with the oxygen atom in the vertex. The length of the segment, i.e. the distance between atoms O and H, is 0.957 Å, and the angle between two segments of this molecule equals 104.5°. There are three types of vibration states for three-atom molecules, and the symmetric vibration of the H_2O molecule, where distances between hydrogen and oxygen atoms are identical in the course of the oscillation, are characterized by the energy $\nu_1 = 3657 \, \mathrm{cm}^{-1}$ (the wavelength is $\lambda_1 = 2.734 \, \mu\mathrm{m}$) for transition from the ground vibration state. In the antisymmetric vibration, hydrogen atoms move in opposite directions with respect to the oxygen atom, and its frequency is equal $\nu_2 = 3756 \, \mathrm{cm}^{-1}$ (the wavelength is $\lambda_2 = 2.662 \, \mu\mathrm{m}$). The lengths of these bonds are not varied at third vibration with change of an angle between segments. Its energy is equal $\nu_3 = 1595 \, \mathrm{cm}^{-1}$ (the wavelength is $\lambda_3 = 6.269 \, \mu\mathrm{m}$).

Note that according to formula (6.3), the intensity of emission of a blackbody whose temperature coincides with the average Earth's temperature $T_E = 288 \, \mathrm{K}$ is $\omega \approx 600 \, \mathrm{cm}^{-1}$ that is less remarkable than vibration frequencies for the water molecules. Hence, radiative vibration transitions of H_2O molecules do not play a part in atmospheric emission because they correspond to far wings of Earth's thermal radiation. These vibrations with respect to an infrared spectrum range are of importance for a wet gas at high temperatures.

It is of importance for rotation radiative transitions in H_2O molecules that the dipole moment of a water molecule is not zero and is equal to $1.85 e a_o$ [50, 51], where e is the electron charge, and a_o is the Bohr radius. Therefore, rotation transitions in water molecules are optically active. Rotational constants of the water molecule are $27.9\,\mathrm{cm}^{-1}$, $14.5\,\mathrm{cm}^{-1}$, and $9.3\,\mathrm{cm}^{-1}$, respectively. Comparing these values with typical photon energies in atmospheric thermal emission, one can see that rotation-excited H_2O molecules may be effective atmospheric radiators. Especially water molecules are effective in emission of low-frequency molecules. Taking the number density of water molecules near the Earth's surface $N(H_2O) \approx 4 \times 10^{17}\,\mathrm{cm}^{-3}$, and a typical altitude of location of water molecules $\lambda = 2\,\mathrm{km}$ according to formula (2.5), one can see that emission of water molecules for a given frequency is of importance, if the optical thickness $u_\omega = N(H_2O)\lambda S \sim 1$, i.e. the intensity of spectral lines due to water molecules $S \gg 10^{-23}\,\mathrm{cm}$. According to Fig. 6.5b, this criterion is fulfilled more or less in the total spectrum of atmospheric emission. Hence, water molecules are of importance in thermal emission of the atmosphere. According to Fig. 6.11, this contribution is approximately 50% under real conditions.

Emission of water microdroplets which form clouds is of importance for atmospheric emission, and, therefore, parameters of thermal atmospheric radiation, in turn, allow us to determine some of the average parameters of clouds. We below consider absorption of a long electromagnetic wave by a small water microdroplet within the framework of the Mie theory [52] which considers scattering of an electromagnetic wave on spherical particles with a sharp boundary. In this case, parameters of scattering follow from sewing of electromagnetic wave parameters at the particle boundary [53–55]. In this consideration, we account for the nature of liquid water that formed a microdroplet. Liquid water is a weak electrolyte that includes H_3O^+ and OH^- ions inside it. Hence, interaction of infrared radiation with water in the region of wavelengths that determines thermal radiation of the atmosphere corresponds to its interaction with dipole momenta created by these positively and negatively charged ions inside the water.

Considering scattering of electromagnetic waves on water droplets within the framework of the Mie theory [52] in the standard method [53–55], we assume the magnetic field of the wave to be relatively small because of a small water conductivity. Indeed, for the stationary electric field and the temperature of liquid water of $25\,^\circ\mathrm{C}$, its specific resistivity is equal to $18\,\mathrm{M\Omega \cdot cm}$ [56, 57] that corresponds to the conductivity $\Sigma = 5 \times 10^6\,\mathrm{s}^{-1}$. Since for the infrared spectrum range we deal with frequencies $\omega \sim 10^{14}\,\mathrm{s}^{-1}$, the criterion

$$\Sigma \ll \omega \tag{6.40}$$

holds true. This criterion allows one to neglect the wave magnetic field in the course of scattering of electromagnetic wave on a water microdroplet.

Let us apply the Mie theory for a water droplet of a small radius r

$$r \ll \lambda, \tag{6.41}$$

where λ is the wavelength of an electromagnet wave. Under the criterion (6.41), the absorption cross section σ_{abs} for a spherical drop and the scattering cross section σ_{sc} of an electromagnetic wave depend on the drop radius as [58]

$$\sigma_{abs} \sim r^3, \ \sigma_{sc} \sim r^6 \tag{6.42}$$

Taking an incident electromagnet wave in the form of a plane wave, we have the Maxwell equations inside the drop in the form

$$curl\mathbf{E} = -\frac{1}{c}\frac{\partial \mathbf{H}}{\partial t}, \ curl\mathbf{H} = \frac{4\pi}{c}\mathbf{j} - \frac{1}{c}\frac{\partial \mathbf{E}}{\partial t},$$

where $\mathbf{E}$ and $\mathbf{H}$ are the electric and magnetic fields of the electromagnetic wave, $\mathbf{j}$ is the density of the electric current produced by the action of the electromagnetic wave, and c is the light velocity. Applying the *curl* operator to the first equation and the operator $-(1/c)(\partial/\partial t)$ to the second equation, and then eliminating the magnetic field from the resulting equation for the electric field strength, one can obtain

$$\nabla div\mathbf{E} - \Delta\mathbf{E} + \frac{4\pi}{c^2}\frac{\partial \mathbf{j}}{\partial t} - \frac{1}{c^2}\frac{\partial^2 \mathbf{E}}{\partial t^2} = 0 \tag{6.43}$$

We now simplify the Maxwell equation (6.43) under conditions related to a water microdroplet. Since electric charges are absent inside water, this gives $div\mathbf{E} = 0$, and the first term is zero. Substituting the expression $\mathbf{j} = \Sigma\mathbf{E}$ into third term of equation (6.43), one can estimate this term as $\sim \Sigma\omega\mathbf{E}/c^2$. Because of the criterion (6.40), this term for a water drop is small compared with the last term of equation (6.43) that is estimated as $\omega^2\mathbf{E}/c^2 \sim \mathbf{E}/\lambda^2$. In turn, the second term of equation (6.43) is estimated as $\sim \mathbf{E}/r^2$ and according to criterion (6.41) it exceeds significantly the last term. In this manner, the Maxwell equation (6.43) under these conditions is reduced to the form

$$\Delta\mathbf{E} = 0 \tag{6.44}$$

This equation gives that the amplitude of the electric field of an electromagnet wave inside a water droplet is described by the stationary equation for an electric field in a matter consisting of a droplet and environment. The magnetic field of an electromagnet wave is estimated in this case inside a water droplet as $H \sim E \cdot \Sigma/\omega$ [58], i.e. one can ignore the magnetic field in this case. Thus, the distribution of the electric field strength inside a droplet for an electromagnet wave is similar to that for a stationary electric field in this matter.

Let us determine now the power which results from interaction between an electromagnet wave and a water droplet inside which dipole moment $\mathbf{D}$ is induced by the electric field of an electromagnetic wave. Since the interaction potential between an electromagnetic wave and the induced dipole moment of a droplet is equal $-\mathbf{ED}$, the power P absorbed by the droplet is equal

$$P = -\left\langle \mathbf{E}\frac{d\mathbf{D}}{dt} \right\rangle, \tag{6.45}$$

where brackets mean averaging over a large time compared to the period of wave oscillations. Taking the electric field strength of a monochromatic electromagnet wave in the form

$$\mathbf{E} = \mathbf{E}_o e^{i\omega t} + \mathbf{E}_o^* e^{-i\omega t},$$

one can obtain for the induced dipole moment

$$\mathbf{D} = \alpha(\omega)\mathbf{E}_o e^{i\omega t} + \alpha^*(\omega)\mathbf{E}_o^* e^{-i\omega t},$$

where $\alpha(\omega)$ is the drop polarizability. This gives for the absorbed power at a given frequency

$$P = i\omega |E_o|^2 \left[\alpha^*(\omega) - \alpha(\omega)\right]$$

According to the definition of the absorption cross section $\sigma_{abs}(\omega)$ of an electromagnetic wave by a droplet, we introduce this quantity as the ratio of the power P absorbed by a droplet to the radiative flux $c |E_o|^2 / (2\pi)$. As a result, we obtain

$$\sigma_{abs} = \frac{4\pi\omega}{c} \cdot Im \; \alpha(\omega) \tag{6.46}$$

Next, one can express now the absorption cross section σ_{abs} for a water microdroplet through the dielectric constant $\varepsilon(\omega)$ of bulk water. For this goal, it is necessary to connect the droplet polarization $\alpha(\omega)$ with the dielectric constant $\varepsilon(\omega)$ of bulk water. Using the continuity of the electric potential $\varphi(R)$ of an electromagnet wave as a function of a distance R from the droplet center, we have at the droplet boundary

$$\varepsilon \cdot \frac{\partial\varphi}{\partial R}(R \to r - 0) = \frac{\partial\varphi}{\partial R}(R \to r + 0), \tag{6.47}$$

where ε is the dielectric constant of bulk water. The electric potential in a vacuum outside the droplet is given by

$$\varphi = -\mathbf{ER} + \frac{\mathbf{DR}}{R^3},$$

where the first term is determined by an external electric field located in a vacuum, whereas the second term is induced by this field inside the droplet. One can represent the electric potential inside the droplet as a solution of the Poisson equation $\Delta\varphi = 0$ with accounting for its finiteness in the form

$$\varphi = A\mathbf{ER},$$

where A is a numerical coefficient which may be determined from the continuity condition (6.47) for the electric potential. From this we obtain

$$A = \frac{3\varepsilon}{\varepsilon + 2}, \quad \alpha = \frac{\varepsilon - 1}{\epsilon + 2} r^3 \tag{6.48}$$

In particular, one can use these relations for the case $\varepsilon = 1$ for scattering of an electromagnetic wave in a vacuum, where we obtain from formula (6.48) $A = 1$, $\alpha = 0$. According to formula (6.46), the absorption cross section in this case is $\sigma_{abs} = 0$. Next, on the basis of formula (6.46), one can express the absorption cross section for a dielectric droplet through the dielectric constant as [58, 59]

$$\sigma_{abs} = \frac{12\pi\omega r^3}{c} \cdot \frac{\varepsilon''}{(\varepsilon' + 2)^2 + (\varepsilon'')^2}, \tag{6.49}$$

where the dielectric constant $\varepsilon(\omega)$ at a given frequency is represented in the form

$$\varepsilon = \varepsilon'(\omega) + i\varepsilon''(\omega),$$

It is convenient to express the absorption cross section (6.49) through parameters of a plane electromagnet wave. Then the dependence of an electric field strength for an electromagnet wave on time t and coordinate $\mathbf{R}$ may be represented as

$$\mathbf{E} = \mathbf{E}_o \exp(-i\omega t + i\mathbf{k}\mathbf{R}),$$

where $\mathbf{E}_o$ does not depend on time and coordinates. The wave vector k of this wave is given by

$$k = \sqrt{\varepsilon} \cdot \frac{\omega}{c} = (n + i\kappa)\frac{\omega}{c} \tag{6.50}$$

This allows one to express the dielectric constant in formula (6.46) through the refractive index of bulk water. Taking the refractive coefficient in the complex form $n + i\kappa$ with κ- the extinction coefficient, one can represent the connection between the refractive coefficient of a bulk matter with its dielectric constant as [58, 60]

$$n^2 = \frac{\sqrt{(\varepsilon')^2 + (\varepsilon'')^2} + \varepsilon'}{2}, \quad \kappa^2 = \frac{\sqrt{(\varepsilon')^2 + (\varepsilon'')^2} - \varepsilon'}{2} \tag{6.51}$$

The inverse connection has the form

$$\varepsilon' = n^2 - \kappa^2, \quad \varepsilon'' = 2n\kappa \tag{6.52}$$

Correspondingly, the attenuation coefficient k which is represented in Fig. 6.14 for the infrared spectrum range is expressed through the extinction coefficient κ as

$$k_\omega = \frac{4\pi\kappa}{\lambda}, \tag{6.53}$$

Fig. 6.13 Refractive index (n) and the extinction coefficient (κ) for liquid water in the infrared spectral range according to [61]

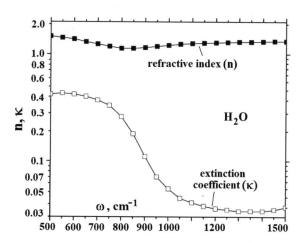

where λ is the wavelength. If a weak signal penetrates from a vacuum into a uniform matter through a flat boundary, according to the Beer–Lambert law [18, 19] its intensity decreases with an increase of a distance z from the boundary as $\exp(-k_\omega z) = \exp(-z/\delta)$, where δ is the penetration depth. Correspondingly, the absorption cross section is expressed through the refractive index and extinction coefficient as

$$\sigma_\omega = \frac{12\pi\omega r^3}{c} \cdot \frac{2n\kappa}{(n^2 + \kappa^2 + 2)^2} \tag{6.54}$$

Thus, formulas (6.49) and (6.53) allow one to determine the absorption cross section of an electromagnet wave at a given frequency ω by a water microdroplet (6.41) on the basis of the dielectric constant $\varepsilon'(\omega)$ and $\varepsilon''(\omega)$ or the refractive index $n(\omega)$ and $\kappa(\omega)$.

Using information about the above electric parameters of liquid water in the infrared spectrum range [61–69], one can determine the absorption cross section by a liquid microdroplet under the criterion (6.41). Figure 6.13 contains the frequency dependence for components of the refractive index in the infrared spectrum range which is responsible for infrared emission and absorption of atmospheric water microdroplets.

This information about the electric parameters of liquid water allows us to determine the character of propagation of an electromagnetic wave through liquid water. Figure 6.14 contains the attenuation coefficient for an electromagnet wave which propagates through liquid water depending on the wave frequency. As it follows from this, the penetration depth δ varies in the frequency range under consideration from 3 μm in the frequency range (600–800) cm^{-1} up to 20 μm in the frequency range (1000–1500) cm^{-1}.

Evidently, radiative transitions in liquid water for an infrared spectrum range related to thermal radiation of the atmosphere are connected with the atomic structure of liquid water [70–73]. Roughly, Fig. 6.14 demonstrates the resonant character of

Fig. 6.14 Attenuation
coefficient for liquid water in
the infrared spectrum range
according to [61]

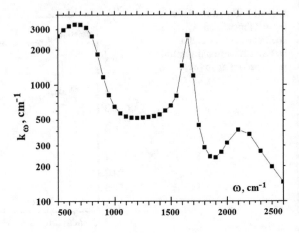

Fig. 6.15 Absorption
coefficient of liquid water
under normal conditions [74,
75]. Arrows indicate the
visible and infrared spectrum
ranges, and the cross refers
to the measurement [76] for
the cirrus-cumulus cloud

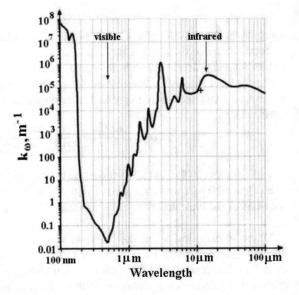

interaction between an electromagnet wave and liquid water. But these resonances
do not corresponds to vibration transitions inside a water molecules which exceed
remarkably these resonant energies. Evidently, they result from hindered rotation of
ions H_3O^+ and OH^- located in liquid water as a weak electrolyte. But interaction
of rotation of radicals with an environment inside the liquid leads to broadening
and mixing of resonances. Nevertheless, existing of the above resonances leads to a
strong absorption of an electromagnet wave.

One can expand the range of frequencies for absorption of an electromagnet wave
in liquid water as it is given in Fig. 6.15, where along the infrared spectrum range,
the visible one is represented. In the scale of this Figure, the penetration depth in

liquid water varies from of the order of 5 m for the visible spectrum range up to of the order of 10 μm for the infrared spectrum range. Applying this to water droplets, one can obtain from this that microdroplets are opaque for infrared photons and are transparent for visible ones.

From this, the Twomey effect [77, 78] follows according to which clouds consisting of water microdroplets are transparent for visible light and are opaque for infrared radiation. But if clouds absorb admixture molecules or dust of a small concentration (~0.1% of the water mass), clouds become visible. This explains an observed phenomenon, when clouds suddenly flare up among a clear sky. In reality, clouds consisting of water microdroplets arise below an observation time, but they consisted of pure water. Absorption of an admixture makes them visible.

Let us apply the above analysis for determination of the absorption cross section of electromagnet wave by a water microdroplet. In the limit of small droplets in accordance with the criterion (6.41), the absorption cross section is given by formula (6.49) with using the above electric parameters for liquid water. Accounting for a strong interaction of an electromagnetic wave with liquid water, one can represent the absorption cross section for a large droplet as

$$\sigma_{abs} = \pi r^2, \ r \gg \lambda \tag{6.55}$$

which corresponds to a blackbody model for the absorbing droplet. Combining formulas (6.49) and (6.55), one can represent the absorption cross section in a wide range of droplet sizes

$$\sigma_{abs} = \frac{\pi r^2}{1 + C(\lambda)\frac{(\omega)}{r}} \tag{6.56}$$

Comparing formulas (6.56) and (6.49), one can represent the expression for the parameter $C(\omega)$ as

$$C(\omega) = \frac{(\varepsilon' + 2)^2 + (\varepsilon'')^2}{24\pi\varepsilon'} = \frac{(n^2 + \kappa^2 + 2)^2 + 4n^2\kappa^2}{48\pi n\kappa} \tag{6.57}$$

Figure 6.16 contains the frequency dependencies for parameters $C(\omega)$ and $\lambda C(\omega)$ in accordance with formula (6.57). Note that formula (6.55) holds true in the limit $r \gg \lambda C(\omega)$. We give also in Fig. 6.17 the frequency dependence of the absorption cross sections of the electromagnet wave by water microdroplets of different radii according to formulas (6.56) and (6.57).

6.2.2 Clouds as Absorbers

We now consider absorption of infrared radiation by clouds consisting of water microdroplets. If we assume clouds to be consisted of microdroplets of an identical size, the measure of photon absorption is the ratio of the absorption cross section

Fig. 6.16 Frequency
dependence for the
parameter $C(\omega)$ (**a**)
according to formula (6.57)
and such a dependence for
the parameter $\lambda C(\omega)$ (**b**)

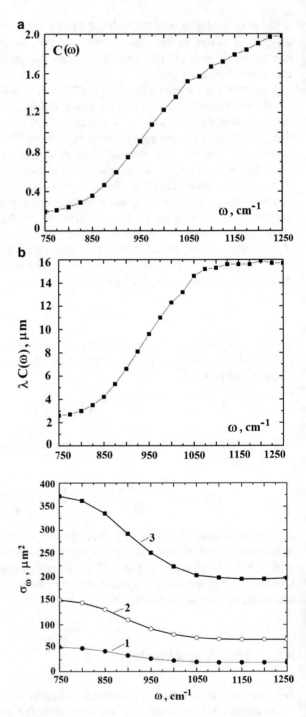

Fig. 6.17 Absorption cross
section of a water
microdroplet according to
formula (6.56) for different
radii r: 1—$r = 5\,\mu m$,
2—$r = 8\,\mu m$,
3—$r = 12\,\mu m$ [29]

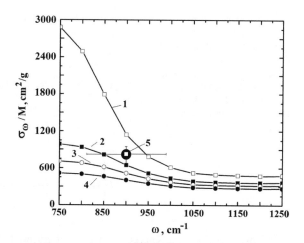

Fig. 6.18 Specific absorption cross section by a liquid water droplet according to formulas (6.56), (6.57), if a cloud consists of droplets of an identical indicated radius r: 1—small radius, 2—$r = 5\,\mu m$, 3—$r = 8\,\mu m$, 4—$r = 12\,\mu m$, 5—experiment [76] for stratocumulus clouds

to the specific mass of water inside microdroplets. Let us determine first this value for an individual microdroplet of a radius R whose mass is $M = 4\pi r^3 \rho/3$, where $\rho = 1\,g/cm^3$ is the mass density of liquid water. Then from formula (6.56), one can find the ratio of the absorption cross section σ_ω to the mass

$$\frac{\sigma_\omega}{M} = \frac{3}{4\rho[r + \lambda C(\omega)]} \tag{6.58}$$

Figure 6.18 represents the frequency dependence for the specific absorption cross section of the electromagnet wave by water microdroplets in the infrared range of frequencies under consideration. If the water mass per column of atmosphere that is contained in microdroplets is ϱ, then the optical thickness of the atmosphere with respect to water microdroplets is

$$u_\omega = \varrho \frac{\sigma_{abs}(\omega)}{M} \tag{6.59}$$

Note that Fig. 6.18 contains the result of the experiment [76] for stratocumulus clouds where the water density was in the range $(0.02-0.3)g/m^3$ and the wavelength of radiation was $\lambda = (10-12)\,\mu m$. According to the indicated measurements, the average specific cross section was $765\,m^2/g$, but the range of location of this value was $(700-1000)\,m^2/g$. From comparing this experimental result with formula (6.58) and its parameters given in Fig. 6.19, the radius of microdroplets which constitute the stratocumulus cloud in this experiment is equal

$$r = (4 \pm 1)\,\mu m \tag{6.60}$$

As is seen, microdrops of stratocumulus clouds are less than those of cumulus clouds with the average radius $r = 8\,\mu m$ according to formula (3.1).

Fig. 6.19 Radiative flux
which is formed at the
Earth's surface and passes
through the atmosphere as a
function of the water mass in
clouds per unit atmospheric
column. Clouds consist of
identical microdroplets of an
indicated radius

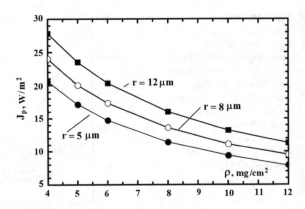

We now analyze the character of pass of infrared radiation through the atmosphere. According to the energetic balance of the Earth given in Fig. 2.15b, the average radiative flux of $20\,W/m^2$ passes through the atmosphere. This flux is created at the beginning by the Earth's surface, and this means that approximately 5% of the power that is emitted by the Earth is not absorbed by the atmosphere. But according to Fig. 6.9, emitted radiation with frequencies below $800\,cm^{-1}$ is absorbed by atmospheric molecules, and hence 13% of the radiative flux with frequencies above $800\,cm^{-1}$ is absorbed by clouds.

Let us determine the radiative flux J_p which is emitted by the Earth's surface and passes through the atmosphere. Following to [23, 29], we introduce the distribution function $f(u_\omega)$ over optical thicknesses u_ω at a given frequency ω that is the probability of a given optical thickness, i.e. the following normalization condition is fulfilled

$$\int_0^\infty f(u_\omega)d\omega = 1$$

This means that the optical thickness of clouds may be different, including the case of an absence of clouds, i.e. a clear sky. For definiteness, we use the simple distribution function [23, 29]

$$f(u_\omega) = \frac{1}{\overline{u_\omega}}\exp\left(-\frac{u_\omega}{\overline{u_\omega}}\right), \tag{6.61}$$

where $\overline{u_\omega}$ is the average optical thickness.

From this, we have for the probability of photon surviving at a given frequency

$$P(\overline{u_\omega}) = \int_0^\infty f(u_\omega)du_\omega \int_0^1 \cos\theta\exp\left(-\frac{u_\omega}{\cos\theta}\right)d\cos\theta = \int_0^1 \frac{\cos\theta}{\overline{u_\omega}+\cos\theta}d\cos\theta \tag{6.62}$$

In limiting cases, we have $P(\overline{u}_\omega) = 1$ at $\ll 1$ and $P(\overline{u}_\omega) = 1/2\overline{u}_\omega$ at $\overline{u}_\omega \gg 1$. On the basis of the limiting cases, one can construct the following expression for the probability of photon surviving

$$P(\overline{u}_\omega) = \frac{1}{1 + 2\overline{u}_\omega} \tag{6.63}$$

In particular, assuming the probability to pass for an emitted photon from the Earth's surface as $P = 13\%$, as it follows from the energetic balance of the atmosphere, one can obtain $\overline{u}_\omega \approx 3$. It should be noted that under used assumptions this is an estimate.

We now evaluate the radiative flux J_p that is emitted by the Earth and is passed through the atmosphere. This flux is given by

$$J_p = \int\limits_0^\infty J_E'(\omega) P(\overline{u}_\omega) d\omega \tag{6.64}$$

Here, $J_E'(\omega)$ is the radiative flux for a given frequency which is emitted by the Earth and reaches the cloud boundary. This flux is represented in Fig. 6.10. Note that though the lower limit of integration is zero, the contribution to this integral from frequencies below $800 \, cm^{-1}$ is small because of absorption by H_2O and CO_2 molecules in this spectrum range.

Evaluation on the basis of formula (6.64) is represented in Fig. 6.22 for some sizes of microdroplets which constitute clouds. Assuming that these sizes are realized in clouds, one can obtain a specific mass of water in a atmospheric column, if the radiative flux from the Earth outside $J_p = 20 \, W/m^2$. Then one can obtain for the specific mass of water microdrops in the atmosphere

$$\varrho = (5 \pm 1) \, mg/cm^2 \tag{6.65}$$

One can see that this value is approximately 0.02% of the mass of atmospheric water. This is small also compared to the water mass contained by cumulus clouds which are responsible for processes of atmospheric electricity.

In addition, Fig. 6.20 contains values of the optical thicknesses for a rare clouds with the specific mass of water in clouds of $5 \, mg/m^2$. This demonstrates also that though above we analyze clouds on the basis of average parameters, values of these parameters under certain conditions may be essential.

Thus, in spite of the roughness of the above analysis with average values of cloud parameters, one can represent a scheme that described the physical picture for atmospheric clouds in radiative processes. Namely, cumulus clouds are responsible for electric processes in the atmosphere and include the main part of condensed atmospheric water whose mass is several percent of the total mass of atmospheric water. But cumulus clouds cover a small part of the Earth's surface, and rare clouds determine radiative properties of the atmosphere. These radiative properties consist

Fig. 6.20 Frequency
dependence for the optical
thickness in the infrared
spectrum range for clouds
consisting of water
microdroplets of an indicated
size if the total mass density
of water microdroplets of an
atmospheric column is
5 mg/cm^2 [29]

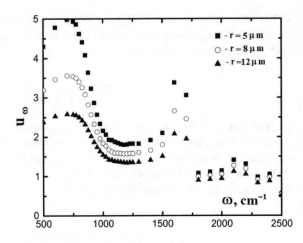

in atmospheric emission toward the Earth and outside, as well as pass of thermal
radiation through the clouds. The mass of these rare clouds, as an example, stratus
ones, includes of the order of 0.1% of the total water mass in the atmosphere.

6.2.3   Absorption of RF-radiation by Clouds with Microdroplets

We consider above the interaction of electromagnet waves with water microdroplets
of clouds being guided by a frequency range which is of importance for thermal
emission of the atmosphere. But for other problems of this type, the interaction of
electromagnet waves of other frequencies is of importance. In particular, propagation of radio-frequency radiation through clouds [79–81] requires the understanding
of these electromagnetic waves with water microdroplets. We consider below the
interaction of long-wave electromagnetic waves with microdroplets of clouds.

In this case the criterion (6.41) holds true, so that the absorption cross section
of a microdroplet according to formula (6.49) is proportional to r^3 (r is the droplet
radius), i.e. the optical thickness of clouds u_ω is independent of a microdroplet size
and is determined by the expression

$$u_\omega = \varrho F(\omega), \quad F(\omega) = \frac{18\pi}{\rho\lambda} \cdot \frac{\varepsilon''(\omega)}{[\varepsilon'(\omega) + 2]^2 + [\varepsilon''(\omega)]^2}, \tag{6.66}$$

where $\rho = 1\,\text{g/cm}^3$ is the mass density of water, $\lambda = 2\pi c/\omega$ is the wavelength, and
ϱ the water mass in cloud microdroplets for an atmospheric column of unit area
which is expressed in g/cm^2. Figure 6.21 contains the frequency dependence of
the real and imaginery parts of the dielectric constant as a frequency function in the
centimeter range of wavelengths in accordance with data of [82]. We give in Fig. 6.22
the frequency dependence of function $F(\omega)$ which is the element of formula (6.66).

As it follows from Fig. 6.22, the quantity $F(\omega)$ is characterized by a sharp dependence on the frequency $\nu = \omega/2\pi$. Hence, different frequencies may be used for measurement of the optical thickness for cumulus and rare clouds. This implies a scheme of measurement of the optical thickness of the atmosphere, where the antenna sends a pulse signal to the flight which is located over the troposphere and receives or reflects this signal. In the latter case the receiver is located at the Earth's surface. We accounts for that atmospheric molecules which do not absorb this radiation, though there are other mechanisms of absorption in a such frequency range due to nonuniformities in the atmosphere. It should be noted also that the results in Figs. 6.21 and 6.22 correspond to microparticles consisting of pure water. We add also to this that clouds are transparent for signals with larger wavelengths.

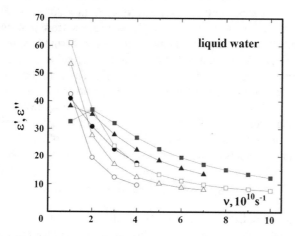

Fig. 6.21 Real and imaginary parts of the dielectric constant for pure liquid water depending on the reciprocal wave number $\nu = \omega/2\pi$. Open signs relate to the real part of the dielectric constant, closed signs correspond to the imaginary part of the dielectric constant. Circles refer to the temperature $0\,^\circ$C, triangles correspond to the temperature $10\,^\circ$C, and squares relate to the temperature $20\,^\circ$C [82]

Fig. 6.22 Frequency dependence of the function $F(\omega)$ which determines the optical thickness of the atmosphere in the indicated frequency range according to formula (6.66). Data [82] are used for the dielectric constant as a frequency function at temperatures $0\,^\circ$C, $10\,^\circ$C and $20\,^\circ$C

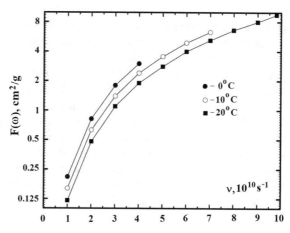

6.3 Peculiarities of Emission of Wet Atmosphere

6.3.1 Emission of Greenhouse Atmospheric Components

Let us formulate one more model that we used in the analysis of thermal emission of the atmosphere. Taking the model "line-by-line" [1] as a basis, we evaluate the radiative flux for each frequency separately, and then summarize radiation fluxes from various frequencies in the total radiative flux. In this consideration [29], the atmosphere is a multicomponent system whose radiators and absorbers are molecules (mostly H_2O and CO_2 molecules) and water microdroplets which constitute clouds. Assuming emission of each radiator as a noncoherent one, we summarize fluxes from each radiator in the radiative flux at a given frequency. In addition, we place clouds at some altitude and in this manner separate clouds in space from molecular radiators.

It is of importance in this consideration the local thermodynamic equilibrium both in atmospheric air, as well as between air and the radiation field. This means that the radiative temperature of each radiator coincides with that of its residence point. In addition, we assume that the optical thickness due to clouds is large, and their boundary is sharp. In this approximation, clouds emit toward the Earth as a blackbody of the temperature T_{cl} of their boundary. Thus, in this consideration the radiative flux to the Earth J_ω is the sum of radiative fluxes of clouds and molecules located in a gap between the Earth's surface and clouds in accordance with formula (6.16)

$$J_\omega = I_\omega(T_\omega)g(u_\omega) + I_\omega(T_{cl})[1 - g(u_\omega)],$$

where $I_\omega(T)$ is the radiative flux of a blackbody with a given temperature, $g(u_\omega)$ is the opaque factor, i.e. it is the probability for a photon of a given frequency to pass the gap between clouds and the Earth without absorption, u_ω is the gap optical thickness, and T_ω is the temperature of emission of molecular components. It is of principle here that we join individual radiators in a whole one, and this is expressed, in particular, in formula (6.37) for the total absorption coefficient due to various molecular components. In this manner, we account for interaction of optically active components where each component is the radiator and absorber simultaneously.

On the basis of this concept and formula (6.16) which corresponds to this concept, one can determine the radiative flux J_ω from the atmosphere to the Earth. In this case, the number densities of main molecular radiators are given by formula (6.27), and the algorithm for determination the radiative parameters of molecular radiators is described in Sect. 6.1.3. But determination of cloud parameters is problematic because condensed water is absent in a motionless atmosphere, and formation of clouds results from fluctuations. One can obtain additional information about clouds as a source of infrared radiation on the basis of the energetic balance of the atmosphere Sect. 2.2.4, so that the total atmospheric radiative flux to the Earth is equal to $327\,W/m^2$, as it follows from Fig. 2.15a. From this, one can find [29] that the altitude of the cloud boundary is $h_{cl} = 4.3\,km$, and the temperature due to clouds is $T_{cl} = 260\,K$.

Fig. 6.23 Atmospheric radiative fluxes toward the Earth due to various components. Radiative fluxes are expressed in W/m^2

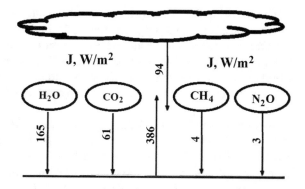

One can use this concept of atmosphere emission for separation of emissions of each component as it is given in Fig. 6.11. In addition, Fig. 6.23 contains the radiative fluxes to the Earth created by each greenhouse component in the frequency range from $700\,cm^{-1}$ up to $780\,cm^{-1}$. In the total, roughly, approximately 50% of the total atmospheric flux toward the Earth is due to atmospheric water molecules, 30% of it is due to clouds, and 20% of the total radiative flux is owing to carbon dioxide molecules. We consider below the problem of separating the contribution of radiating components to the total radiative flux from the atmosphere in detail.

In separating the contributions from various components, we evaluate probabilities that the radiative flux is created by a certain component at a given frequency. We first introduce the probability $\xi_\omega(cl)$ that photons directed to the Earth's surface are created by clouds. According to formula (6.16) we have

$$\xi_\omega(cl) = \frac{I_\omega(T_{cl})[1 - g(u_\omega)]}{I_\omega(T_\omega)g(u_\omega) + I_\omega(T_{cl})[1 - g(u_\omega)]}, \tag{6.67}$$

where T_ω is the radiative temperature of the atmosphere molecular components, and T_{cl} is the cloud radiative temperature. As it follows from Figs. 6.9 and 6.10, clouds give a contribution to the atmosphere radiative flux at high frequencies $\omega > 740\,cm^{-1}$, where $\hbar\omega \gg T_\omega$. At such frequencies $I_\omega \sim \exp(-\hbar\omega/T_\omega)$. Next, we use the approximation (6.18)

$$g(u_\omega) = 1 - \exp(-1.6u_\omega),$$

which holds true in a wide range of frequencies $u_\omega \sim 1$.

Finally, one can obtain from this

$$\xi_\omega = \frac{1}{\exp\left(\frac{\hbar\omega \Delta T}{T_\omega T_{cl}}\right) \cdot [\exp(1.6u_\omega) - 1] + 1}, \tag{6.68}$$

where $\Delta T = (T_\omega + T_{cl})/2$.

One can simplify this expression under conditions where the optical thickness of the atmosphere is not large. At low optical thicknesses u_ω of the gap between the Earth's surface and clouds, we have from equation (6.15) $u(h_\omega) = u_\omega/2$. In particular, the precise solution of equation (6.15) gives $u(h_\omega) = 0.45$ at $u_\omega = 1$, and $u(h_\omega) = 0.38$ at $u_\omega = 0.8$. This allows one to determine the effective altitude h_ω at a given frequency and the radiative temperature of the molecular atmospheric component.

As a result, we obtain the following expression for the altitude h_ω of a layer which is responsible for atmospheric emission at a given frequency. In the limit, where radiation of CO_2 molecules dominate, we have $h_\omega = h_{cl}/2$ at low optical thicknesses u_ω of the molecular components. In other limiting case with H_2O absorbers of the atmosphere, we have from the expression (6.29) $h_\omega = \lambda \ln 2$. Taking $h_{cl} = 3.2$ km, $T_{cl} = 267$ K, one can obtain in limiting cases the effective altitudes h_ω are 1.6 and 1.4 km correspondingly. From this, one can find that the radiative temperature of molecular components for the standard atmosphere ranges from 278 to 279 K. This allows one the expression (6.68) to the form

$$\xi_\omega = \frac{1}{\exp\left(\frac{\omega}{\omega_o}\right) \cdot [\exp(1.6u_\omega) - 1] + 1}, \tag{6.69}$$

where $\omega_o = 2500$ cm^{-1}. Figure 6.24 gives the portion of atmospheric emission depending on the optical thickness of the atmosphere.

As it follows from Fig. 6.24, the portion due to water microdroplets of clouds depends weakly on the frequency. It is convenient to approximate the dependencies of Fig. 6.24 by the expression

$$\xi_\omega = 2.1u_\omega \cdot \exp(-0.92u_\omega) \tag{6.70}$$

One can introduce the probability $P(h)$ that a photon of a given frequency ω is formed at a given altitude h due to emission of H_2O molecules

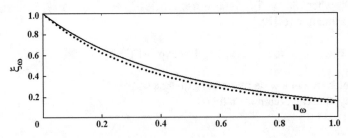

Fig. 6.24 Dependence of the part of the atmospheric emission flux due to water microdroplets of clouds on the optical atmospheric thickness between the Earth's surface and clouds in accordance with formula (6.68). This part of atmospheric emission at the frequency 1200 cm^{-1} is given by a solid curve and at frequency $\omega = 800$ cm^{-1} is represented by a dotted curve

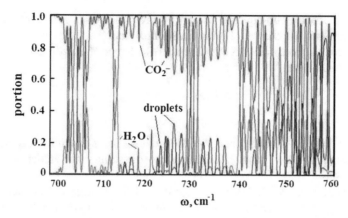

Fig. 6.25 Portion of the radiative atmospheric flux which passes to the Earth's surface due to CO_2 molecules (red), H_2O molecules (green), and clouds (blue) in the frequency range $(700\text{–}760)\,cm^{-1}$ [29]

$$P(h) = \frac{k_\omega(H_2O)\exp\left(-\frac{h}{\lambda}\right)}{k_\omega(H_2O)\exp\left(-\frac{h}{\lambda}\right) + k_\omega(CO_2)\exp\left(-\frac{h}{\Lambda}\right)} \tag{6.71}$$

where $k_\omega(H_2O)$ and $k_\omega(CO_2)$ are the absorption coefficients due to H_2O and CO_2 molecules at the Earth's surface.

Let us denote by $\xi(H_2O)$ the portion of atmospheric emission created by H_2O molecules and by $\xi(H_2O)$ that due to CO_2 molecules. These values may be obtained as a result of an average of the probabilities (6.71) over altitudes. One can introduce the probability $P(h_\omega)$ that a test photon attained the Earth's surface and created by H_2O molecule is equal

$$P(h_\omega) = \langle P(h) \rangle \cdot (1 - \xi_\omega) \tag{6.72}$$

by analogy with the effective altitude of emission for a photon of a given frequency. We then obtain

$$\xi(H_2O) = P(h_\omega) \cdot \xi_\omega, \ \ \xi(H_2O) = [1 - P(h_\omega)] \cdot \xi_\omega \tag{6.73}$$

Figure 6.25 contains the portions of the incident radiative flux to the Earth's surface due to various greenhouse components according to formulas (6.70) and (6.73).

6.3.2   Radiative Flux from Varied Atmosphere

The scheme of atmospheric emission (6.16) allows one to determine the contribution of each greenhouse component to the radiative flux from the atmosphere to the Earth, as well as a change of the flux as a result of change of the atmosphere composition.

We use this scheme subsequently by applying it to radiative flux at each frequency. Let us introduce the change of the radiative flux from the atmosphere to the Earth $\Delta J(\omega)$ due to a given component as a result of a change of atmosphere composition under consideration

$$\Delta J(\omega) = \int_0^\omega \Delta j_\omega d\omega, \tag{6.74}$$

where $\Delta j_\omega = j'_\omega - j_\omega$ is the change of the radiative flux per unit frequency, and j_ω, j'_ω are the partial fluxes at a given frequency for the final and initial atmospheric composition.

Let us introduce $\Delta J(H_2O)$, $\Delta J(CO_2)$, and ΔJ_{cl} the change of the radiative fluxes per unit frequency at a given change of the atmospheric composition due to emission of H_2O, CO_2 molecules and owing to water microdroplets correspondingly. We denote also ΔJ_t—the change of the total radiative flux toward the Earth in this case. Evidently, we have the following relations for these values

$$\Delta J_t = \Delta J(CO_2) + \Delta J(H_2O) + \Delta J_{cl} \tag{6.75}$$

For definiteness, we consider below the change of these radiative fluxes as a result of doubling of the concentration for carbon dioxide molecules. The changes of radiative fluxes for indicated frequency ranges are represented in Table 6.1 [29]. From this Table, it follows that an increase of the fluxes due to emission of CO_2 molecules is accompanied by a decrease of those owing to water molecules and water microdroplets. Therefore, the total change of the radiative flux ΔJ_t in a given frequency range is less remarkable than that due to emission of CO_2 molecules.

We also note an surprisingly large contribution of the "laser" range of frequencies $(950–1100)\,\mathrm{cm}^{-1}$ to the total change of the radiative flux. Indeed, in this frequency range, the atmospheric optical thickness is small, and this range gives a contribution approximately 2% to the radiative flux created by CO_2 molecules, but the contribution to the change of the radiative flux is about one third. This also testifies about the importance of numerical evaluations. Figure 6.26 exhibits the character of change of radiative fluxes according to formula (6.74) as the frequency varies. We define the radiative fluxes J_c, j_c, J_t which are the radiative fluxes due to carbon dioxide molecules, its change at doubling of the concentration of CO_2 molecules, and ΔJ_t is the change of the total radiative flux, so that J'_c and J'_t are those values at the doubling concentration of CO_2 molecules. Hence, these radiative fluxes are defined as

$$J_c = \int_0^\omega J_\omega(CO_2)d\omega, \ \ \Delta j_c = \int_0^\omega [J'_c - J_c]d\omega, \ \ \Delta J_t = \int [J'_t(\omega) - J_t(\omega)]d\omega \tag{6.76}$$

As it follows from Fig. 6.26, the radiative flux J_c toward the Earth's surface due to CO_2 molecules is created mostly inside the absorption band of this molecule between $580\,\mathrm{cm}^{-1}$ and $760\,\mathrm{cm}^{-1}$ which is responsible for thermal radiation due to

Table 6.1 Changes of radiative fluxes based on the model of standard atmosphere toward the Earth as a result of doubling of the concentration of CO_2 molecules in the infrared spectrum range [29]. The frequency ranges are given in cm^{-1}, and changes of radiative fluxes are expressed in W/m^2

Frequency range, $\Delta\omega$	$\Delta J(CO_2)$	$\Delta J(H_2O)$	ΔJ_{cl}	ΔJ_t
580–600	0.96	−0.89	−0.04	0.03
600–620	0.81	−0.74	−0.03	0.04
620–640	0.63	−0.61	0	0.02
640–660	0.15	−0.14	0	0.01
660–680	0.18	−0.18	0	0
680–700	0.21	−0.20	0	0.01
700–720	0.12	−0.03	−0.03	0.06
720–740	0.64	−0.05	−0.39	0.20
740–760	1.07	−0.10	−0.68	0.29
760–780	0.56	−0.02	−0.40	0.14
780–800	0.25	−0.02	−0.17	0.06
800–850	0.15	−0.03	−0.08	0.04
900–950	0.20	0	−0.16	0.04
950–1000	0.76	−0.01	−0.53	0.22
1000–1050	0.18	0	−0.13	0.05
1050–1100	0.37	0	−0.26	0.11
Total	7.24	−3.02	−2.90	1.32

atmospheric carbon dioxide. The change of this radiative flux ΔJ_c as a result of carbon dioxide doubling in the atmosphere is formed both inside the main absorption band for the CO_2 molecule and near the violet boundary of this absorption band. Some contributions to this change follow also from the "laser" range of frequencies between $950\,cm^{-1}$ and $1000\,cm^{-1}$ with a small absorption coefficient (for example, Figs. 6.9 and 6.10). It should be noted that inside the absorption band between $580\,cm^{-1}$ and $760\,cm^{-1}$ the change of the flux due to CO_2 molecules according to more rough evaluations [83–85] is equal approximately $4\,W/m^2$. We now are based on the HITRAN spectroscopic data and hence these evaluations are more accurate.

As it follows from Fig. 6.26, the change of the total radiative flux ΔJ_t is created mostly at the violet boundary of the absorption band of carbon dioxide molecules as a result of doubling of the carbon dioxide concentration, so that approximately 60% of the change of the total radiative flux ΔJ_t relates to this spectral range. In addition, the contribution of other boundaries of the absorption band and inside it is small because of strong absorption by water molecules there. An additional contribution to ΔJ_t results from the "laser" range of frequencies where absorption by CO_2 molecules is small. Thus, one can explain the behavior of the changes of radiative fluxes, but their numerical values cannot be predicted.

Fig. 6.26 Changes of radiative fluxes toward the Earth as a result of doubling of the carbon dioxide concentration in an indicated frequency range are defined by formula (6.74), so that J_c is the radiative flux due to CO_2 molecules, Δj_c is the change of this flux at doubling of the carbon dioxide concentration in the atmosphere, and ΔJ_t is the change of the total radiative flux toward the Earth's surface

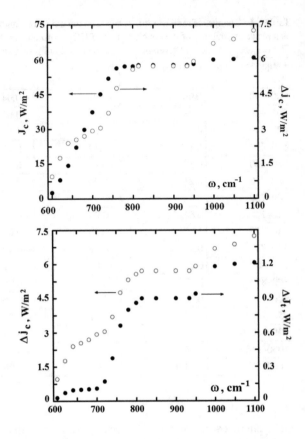

Table 6.1 represents changes of the atmospheric radiative flux in an indicated spectrum range due to doubling of the carbon dioxide concentration. In the case of the concentration change for CO_2 molecules, the change of the total radiative flux proceeds mostly in the frequency range $(700\text{–}800)\,\mathrm{cm}^{-1}$, i.e. at the boundary of absorption band of these molecules. From data of Table 6.2, it follows also that the relative change of concentration of CO_2 molecules influences stronger on the evolution of the total radiative flux than that due to H_2O molecules, though the concentration of water molecules in the standard atmosphere exceeds remarkably that of carbon dioxide molecules. This is explained by strong absorption by CO_2 molecules at the boundary of the absorption band due to water molecules at frequencies $(600\text{–}700)\,\mathrm{cm}^{-1}$. For water molecules the main contribution to the derivative of the flux change over the atmospheric concentration of optically active molecules is determined by a range $(500\text{–}600)\,\mathrm{cm}^{-1}$ that is outside the absorption band of CO_2 molecules.

We now consider a change of the radiative flux from the atmosphere to the Earth as a result of a change of the amount of atmospheric water. Figure 6.27 gives the change of the radiative temperature of the atmosphere in the frequency range of

Table 6.2 The change of the radiative flux ΔJ from the atmosphere to the Earth if the average concentration of water molecules varies from its contemporary value denoted by $c(H_2O)$ up to an indicated one, and the contemporary concentration of atmospheric CO_2 molecules is not varied. The radiative flux changes are expressed in W/m^2 [29]

Frequency range	$0.5c(H_2O)$	$0.9c(H_2O)$	$1.1c(H_2O)$	$2c(H_2O)$
300–400	−0.02	0	0	0.01
400–500	−0.23	−0.02	0.02	0.10
500–580	−0.52	−0.07	0.06	0.29
580–700	−0.14	−0.02	0.02	0.11
700–800	−0.09	−0.01	0.01	0.04
800–1100	−0.07	−0.01	0.01	0.10
Total	−1.07	−0.13	0.12	0.65

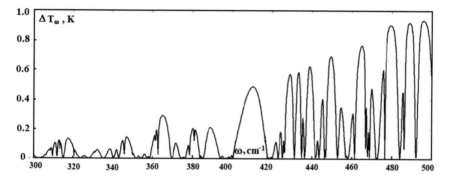

Fig. 6.27 Decrease of the radiative temperature as a result of removal of one-half amount of water under conditions of the model of standard atmosphere [29]

strong atmospheric absorption if the amount of atmospheric water decreases twice. Absorption by water molecules is increased at a decreasing frequency. One can see that the change of the radiative temperature decreases with a decreasing frequency, that is, in the course of an increase of absorption.

Table 6.2 contains the change of the atmospheric radiative flux in an indicated spectrum range and at given changes of the concentration of atmospheric water. As is seen, in spite of larger concentration of water molecules in the standard atmosphere compared with that of carbon dioxide molecules, their contribution to a change of the radiative flux is less than that in the case of CO_2 molecules. This is explained by strong absorption by CO_2 molecules at the boundary of the absorption band due to water molecules at frequencies $(600-700)\,\mathrm{cm}^{-1}$. The main contribution to the derivative of the flux change over the atmospheric concentration of optically active molecules is determined by a range $(500-600)\,\mathrm{cm}^{-1}$ for water molecules which are outside the absorption band of CO_2 molecules. In the same manner, the stronger change of the radiative flux due to a change of the concentration of CO_2 molecules corresponds to the frequency range $(700–800)\,cm^{-1}$ which is at the boundary of absorption band of these molecules.

As it follows from data of Tables 6.2 and Fig. 6.27, the dependence of the radiative flux change on the concentration variation is not linear and not symmetric with respect to an increase and decrease of the molecule concentration. Nevertheless, one can evaluate with a restricted accuracy that is estimated as $(10–20)\%$, the corresponding derivatives

$$\frac{dJ_\downarrow}{d\ln c(CO_2)} = 1.8\frac{W}{m^2}, \quad \frac{dJ_\downarrow}{d\ln c(H_2O)} = 1.3\frac{W}{m^2} \qquad (6.77)$$

One can see that the radiative flux change due to atmospheric water is less than that for carbon dioxide, though atmospheric water is more important for the atmospheric emission toward the Earth. The reason is connected with screening of radiation of water molecules by carbon dioxide ones at the boundary of the absorption band, whereas just this spectrum range gives the main contribution to the change of the radiative flux due to CO_2 molecules.

As it follows from data of Table 6.2, the dependence of the radiative flux change on the concentration variation is not linear and not symmetric with respect to an increase and decrease of the molecule concentration. At low concentration changes, we have [29]

$$\frac{dJ_t}{d\ln c(CO_2)} = 1.8\frac{W}{m^2}, \quad \frac{dJ_t}{d\ln c(H_2O)} = 1.3\frac{W}{m^2} \qquad (6.78)$$

In accordance with the above conclusion, we obtain that the radiative flux change due to atmospheric water is less than that for carbon dioxide, though the concentration of water molecules is large compared with that of carbon dioxide molecules. This was explained above due to character of spectra of these molecules.

6.3.3 Greenhouse Phenomenon of Atmosphere

The greenhouse effect or the greenhouse phenomenon of the atmosphere is the variation of the global temperature (the average temperature of the Earth's surface) as a result of the change of the atmosphere composition. Above we evaluate the change of the radiative flux created in the atmosphere and directed to the Earth's surface as a result of the change of concentrations of optically active components in atmospheric air. We describe below the scheme to transition from the change of the radiative flux to that for the global temperature. This standard scheme is represented in Fig. 6.28 and demonstrates the connection between the change of the radiative flux which is absorbed by the Earth and the change of the global temperature. In this case, an increase of the energy flux due to the change of the atmosphere composition is compensated by energy fluxes which are caused by the change of the global temperature.

In this analysis, we introduce the climate sensitivity S [83, 86] as the ratio of the global temperature variation ΔT to the change ΔJ_t of the total radiative flux toward

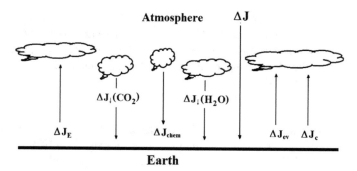

Fig. 6.28 Energy fluxes resulted from an additional energy flux to the Earth's surface. Here, ΔJ is a small external flux of the energy that induces indicated energy fluxes. These fluxes are proportional to the temperature change and use the following notations: ΔJ_E is the radiative flux change from the Earth's surface, the fluxes changes for atmospheric radiation are $\Delta J_{\downarrow}(CO_2)$ due to carbon dioxide molecules and $\Delta J_{\downarrow}(H_2O)$ due to water molecules correspondingly, ΔJ_{chem} is the change of the radiative flux due to the change in the number density of water and carbon dioxide molecules in the atmosphere because of the equilibrium between the Earth and atmosphere, ΔJ_{ev} is the change of the energy flux due to water evaporation from the Earth's surface, and ΔJ_c is that due to air convection [29]

the Earth, as well as a reciprocal quantity, the radiative forcing F, that is defined as the ratio of the change of the total radiative flux absorbed by the Earth in accordance with relations

$$S = \frac{\Delta T}{\Delta J_t}, \quad F = \frac{1}{S} = \frac{\Delta J_t}{\Delta T} \tag{6.79}$$

Since the change of the total flux to the Earth is the sum of partial ones, the radiative forcing is the sum of those for all the channels.

The climate sensitivity is sensitive to processes which are included in consideration. For example, in Fig. 6.28 an interaction between the atmosphere and ocean is absent, whereas this interaction may be taken into account in different ways. Hence, the accuracy of determination of this value is low. For example, this value determined in [86] for past events and different geographical points lies within the limits between 0.3 and $1.9(m^2 \cdot K/W)$. The same study for other data in past [87] gives for the corridor of this value from 0.25 up to $0.79(m^2 \cdot K/W)$. Nevertheless, we use results of some evaluations of this value related to different conditions. Namely, the climate sensitivity is $0.55(m^2 \cdot K/W)$ according to [83], $0.64(m^2 \cdot K/W)$ according to [88], $0.49(m^2 \cdot K/W)$ according to [89], and $0.42(m^2 \cdot K/W)$ according to [85]. Statistical averaging of these data gives [29]

$$S \approx (0.5 \pm 0.1)\frac{m^2 \cdot K}{W} \tag{6.80}$$

In reality, the accuracy of this value is larger, and we estimate as 50%.

Let us define the Equilibrium Climate Sensitivity (ECS), that is, the change of the global temperature as a result of the doubling of the concentration of CO_2 molecules [90]. This quantity may be used as a characteristic of the influence of the atmospheric composition on the global temperature. On the basis of data of Table 6.1, one can obtain for this value [29]

$$ECS = (0.6 \pm 0.3)\,°C \tag{6.81}$$

In spite of the bad accuracy, this value contradicts to the conclusion of the Intergovernmental Panel on Climate Change [91], according to which ECS ranges from 1.5 °C up to 4.5 °C. This corresponds to statistical averaging of calculation [92–99] according to which

$$ECS = (3.0 \pm 1.5)\,°C \tag{6.82}$$

It is of principle to understand the reason for this discrepancy that follows from data of Table 6.1. Indeed, basing on experience of the first analysis of the greenhouse effect due to atmospheric CO_2 molecules, climatological models were ignored by interaction of greenhouse components. In other words, it ignored a decrease of the radiative flux due water molecules and water droplets after an increase in the amount of atmospheric CO_2 molecules. This means that instead of the change of the total radiative flux as a result of doubling of carbon dioxide concentration ΔJ_t was used so that $\Delta J(CO_2)$ due to CO_2 molecules, i.e. according to data of Table 6.1 the value $\Delta J(CO_2) = 7.2\,W/m^2$ was taken in the climatological evaluations instead of $\Delta J_t = 1.3\,W/m^2$. As is seen, this leads to a mistake more than in fife times that corresponds to the difference of results due to formulas (6.81) and (6.82).

In order to understand the reason of this mistake in climatological codes, let us consider the history of evaluation of the greenhouse effect in the atmosphere. Some papers related to development of this problem are collected in the book [100]. Our position is that climatological models do not take into account the overlapping of spectra of CO_2 molecules and atmospheric water, and this leads to the mistake approximately in fife times. This mistake is connected with the history of study of this problem, and we consider it briefly. Swedish scientist S. Arrenius was first who sets this problem at the end of nineteenth century. In the paper of 1896 [101] he asked "Is the temperature of the ground in any way influenced by the presence of heat-absorbing gases in the atmosphere "? The answer to this question was awkward because of the science state at that time. In addition, the goal of the Arrhenius paper was to treat the Langley experiments where the night absorption over the spectrum is measured for solar radiation which was scattered from the moon. For CO_2 molecules the strong absorption was observed near the strongest spectral line of 4.3 μm for this molecule.

For the strongest transitions, Arrhenius can consider the absorption by CO_2 and H_2O molecules independently. Note that the strongest vibration transitions of these molecules give a small contribution to the radiative flux because these transitions correspond to a far tail of the spectrum of atmospheric thermal emission. For example, the atmospheric radiative flux to the Earth due to the resonant transition of 4.3 μm is about $0.5\,W/m^2$ [29] in comparison with the total radiative flux of $327\,W/m^2$.

Subsequently, the problem of atmospheric carbon dioxide was set strictly. For example, that title of the Calendar paper of 1949 [102] was "Can carbon dioxide influence climate"?

It should be noted that the reliability of numerical calculations for the radiative flux of atmospheric emission is determined by information about radiative transitions in atmospheric molecules. The important contribution in this study was made by G.N.Plass and collaborators [93, 103] in fifties. These evaluations were restricted by the spectrum range (12–18) μm which includes basic emission of CO_2 molecules in the real atmosphere, and they are based on the regular (Elsasser) model [104] with a random distribution of spectral lines in the frequency space. In addition, new information for parameters of radiative transitions of CO_2 and H_2O molecules was used in these evaluations. Along with evaluation of ECC which corresponded to formula (6.82), it estimated the influence of H_2O molecules on the global temperature change as a result of doubling of the concentration of atmospheric CO_2 molecules. This effect was estimated as 20%.

One can conclude from this that the wrong position about spectral interaction of atmospheric CO_2 molecules and atmospheric water followed from a lack of information about radiative transitions in atmospheric molecules and atmospheric particles. This detailed information is available now due to the HITRAN database for molecules and other data banks. Evidently, contemporary climatological codes ignore this information and give a wrong result both for the role of CO_2 molecules in the climate change according to formula (6.82) and for the contribution of trace molecules the radiative flux from the atmosphere. We also note that formula (6.81) is based on a careful evaluation [29] of atmospheric radiative fluxes. The analysis [105, 106] shows that more crude models with overlapping of spectrum of CO_2 molecules and other atmospheric radiators lead to ECC values in the limits of formula (6.81).

In addition to the above analysis, we note that the values (6.81) and (6.82) are the changes of the global temperature at the doubling of the concentration of atmospheric CO_2 molecules if other atmospheric parameters are not varied. In reality, one can determine this value for the real atmosphere, and then the concentration of atmospheric CO_2 molecules is only an indicator of the atmosphere state. In the case of the present atmosphere, if we divide the change of the global temperature according to Fig. 2.9 to the change of the concentration of CO_2 molecules in accordance with Fig. 2.14a, then one can obtain formula (2.43)

$$ECC = (2.1 \pm 0.4)\,°C \qquad (6.83)$$

This corresponds to formula (6.82), but in this case other reasons than those connected with CO_2 molecules are responsible for the global temperature change.

One can determine the ECS value in past on the basis of Fig. 2.11. Here the amplitude of the temperature in the course of the glacial epoch is approximately 14 K, and the concentration of CO_2 molecules varies from $c_1 = 180$ ppm up to $c_2 = 280$ ppm during this period. Then we have

$$\mathrm{ECC} = \frac{\Delta T}{\mathrm{d}\ln(c_2/c_1)} = 32\,\mathrm{K} \tag{6.84}$$

This value contains the connection between the intensity of solar radiation and the concentration of CO_2 molecules in the Earth's atmosphere. But in this case the global temperature change is determined by the change of the solar radiation in the Earth's atmosphere due to the change of the Earth-Sun distance in accordance with the Milankovitch theory [107, 108].

6.3.4 Outgoing Atmospheric Radiation

In considering thermal outgoing emission of the atmosphere, we use the same method of evaluation of radiative fluxes as well as that for emission toward the Earth. Analyzing the energetic balance of the Earth and atmosphere represented in Fig. 2.15, one can conclude a large optical thickness of the atmosphere that allows one to separate radiative fluxes toward the Earth and outside. Hence, we will use formula (6.16) for the outgoing radiative flux with taking into account that the contribution to this flux follows from emission of H_2O and CO_2 molecules, as well as by water microdroplets and microparticles.

As early, in this analysis we use as a basis the model of standard atmosphere and at first will be guided by a simple model with averaging the absorption coefficient over frequencies. Then on the basis of formulas (6.23) and (6.24), we have for the effective altitude $h_\uparrow = 6.8\,\mathrm{km}$ with the average radiative temperature $T_\uparrow = 244\,\mathrm{K}$. At this altitude we have $N(H_2O) = 1 \times 10^{16}\,\mathrm{cm}^{-3}$ and $N(CO_2) = 4 \times 10^{15}\,\mathrm{cm}^{-3}$. Next, from the model of standard atmosphere [109], the atmospheric temperature decreases with an altitude increasing with the gradient $\mathrm{d}T/\mathrm{d}h = -6.5\,\mathrm{K/km}$ up to altitude $h = 11\,\mathrm{km}$, and between altitudes of 11 and 22 km the atmospheric temperature is constant $T = 217\,\mathrm{K}$.

In this case similar to atmospheric emission toward the Earth, in the case of outgoing radiation the radiative flux is determined by H_2O and CO_2 molecules at low frequencies and by water droplets or water particles at high frequencies. But since the number density of the above molecules at large altitudes is lower than that near the Earth's surface, overlapping of spectra of these molecules is weak and we ignore this below. As early, emission of CO_2 molecules proceeds in the absorption band between $580\,\mathrm{cm}^{-1}$ and $760\,\mathrm{cm}^{-1}$, while water molecules create radiation at lower frequencies. Figure 6.29 contains the optical thickness of the atmosphere in the range of the absorption band of CO_2 molecules [29] for the model of standard atmosphere.

Note that the problem of emission of outgoing radiation of the atmosphere for clear sky was considered in [111–113] in the same method as it was made for atmospheric emission toward the Earth. These results are represented in Fig. 6.30 for the clear-sky case. This radiation is measured from satellites and this frequency distribution depends on the atmosphere cloudness. In the case of Fig. 6.30, we consider a clear-

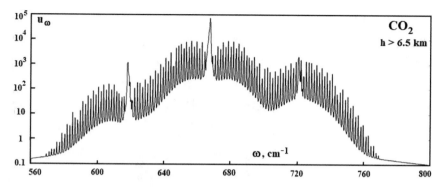

Fig. 6.29 Optical thickness of the atmosphere for outgoing radiation due to radiation of CO_2 molecules above the clouds located at the altitude $h = 6.5$ km

sky atmosphere and the radiative flux in the window of atmosphere transparency that corresponds to the temperature of the Earth's surface. We consider below this problem for standard atmosphere where its energetic balance is given by Fig. 2.15. It is clear that molecular components give a small contribution to the outgoing radiation compared with that toward the Earth. Since clouds give the contribution of approximately 30% to the radiative flux directed to the Earth, their contribution to outgoing radiation is larger.

Usually satellites are located above the stratosphere, and then the stratospheric ozone contributes to atmospheric emission, as it is given in Fig. 6.30. We are restricted by altitudes below 22 km. Since ozone is concentrated in the altitude range between 20 and 35 km, ozone does not contribute to the radiative flux in this consideration in contrast to Fig. 6.30. We start from emission of CO_2 molecules which absorption band ranges from $\omega_1 = 560\,\mathrm{cm}^{-1}$ up to $\omega_2 = 760\,\mathrm{cm}^{-1}$. Taking the concentration of CO_2 molecules in atmospheric air in this range $c = 4 \times 10^{-4}$ with respect to air molecules, one can obtain that practically for all frequencies of the absorption band the effective altitude of radiation ranges between $h = 11$ km and $h = 22$ km with temperature $T_c = 217$ K. Taking the radiative temperature in this frequency range $T_\omega = T_c$, one can determine the radiative flux due to these frequencies

$$\int_{\omega_1}^{\omega_2} I_\omega(T_\omega)g(\omega)d\omega = 24\,\mathrm{W/m^2}, \quad \int_{\omega_1}^{\omega_2} I_\omega(T_\omega)d\omega = 27\,\mathrm{W/m^2}, \qquad (6.85)$$

where $I_\omega(T)$ is the equilibrium radiative flux at frequency ω from a blackbody radiator of a temperature T which is given by the Planck formula (6.3). As it follows from (6.85), the optical thickness of radiation is not small at these frequencies, and the opaque factor $g(\omega) \approx 1$ in this frequency range.

The analysis of atmospheric emission inside the absorption band of CO_2 molecules allows us to formulate the character of radiation. In this range along with emission inside the absorption band, CO_2 molecules screen emission of clouds, so that a small

Fig. 6.30 Outgoing
atmospheric radiative flux
for the case of the fine
atmosphere [110]

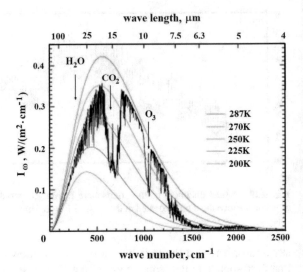

part of cloud emission goes outside. The total radiative flux from this frequency range
is given by

$$J_\uparrow(CO_2) = \int_{\omega_1}^{\omega_2} I_\omega(T_\omega)g(\omega)d\omega + \int_{\omega_1}^{\omega_2} I_\omega(T_{cl})[1-g(\omega)]d\omega = 30\,W/m^2, \quad (6.86)$$

where we take $T_{cl} = 244\,K$.

We assume atmospheric emission at frequencies below $\omega_1 = 580\,cm^{-1}$ to be created by water molecules, whereas at frequencies above ω_2 it is determined by cloud
radiation. In evaluation the radiative flux due to water molecules, we are based on
data of Fig. 6.30 assuming that the radiative flux varies linearly with frequency and
is independent of the cloud temperature. We then obtain for the radiative flux from
this frequency range

$$J_\uparrow(H_2O) = \int_0^{\omega_1} J_\omega d\omega = 98\,W/m^2 \quad (6.87)$$

As a result, one can obtain for the outgoing radiative flux from the standard atmosphere $J_\uparrow$, if it is formed in the troposphere

$$J_\uparrow = J_\uparrow(H_2O) + J_\uparrow(CO_2) + \int_{\omega_2}^\infty I_\omega(T_{cl})d\omega \quad (6.88)$$

As is seen, the total outgoing radiative flux depends on the cloud temperature, i.e.
from an altitude from which clouds radiate. We give in Fig. 6.31 the dependence of

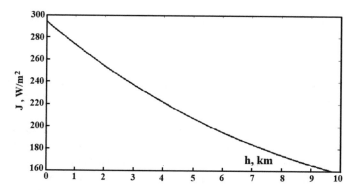

Fig. 6.31 Outgoing radiative flux of the troposphere depending on the altitude h where cloud emission is formed

the effective altitude of cloud radiation for the outgoing radiative flux of standard atmosphere, where the connection between the layer temperature T and its altitude h has the form (see Sect. 2.1.1)

$$T = T_E - \frac{dT}{dh}h, \quad T_E = 288K, \quad \frac{dT}{dh} = 6.5 K/km \qquad (6.89)$$

In particular, on the basis of the energetic balance of the Earth given in Fig. 2.15, the average outgoing radiative flux from the atmosphere is equal to $J_\uparrow = 200\,W/m^2$. This radiative flux and Fig. 6.31 lead to the altitude of cloud emission $h = 5.6\,km$, and the cloud temperature is equal $T_{cl} = 252\,K$ at this altitude.

Let us consider one more element of the Earth energetics given in Fig. 2.15 which corresponds to propagation of the radiative flux through the atmosphere on average. In accordance with Fig. 2.15b, the average radiative flux emitted by the Earth's surface and passed through the atmosphere equals to $J_p = 20\,W/m^2$. In accordance with formula (6.10), one can determine this value for standard atmosphere in the form

$$J_p = \int I_\omega \left[1 - g(u_\omega + u_\omega^{cl}) \right] d\omega, \qquad (6.90)$$

where I_ω is the radiative flux emitted from the Earth's surface, which is a blackbody with the temperature $T_E = 288\,K$, u_ω is the optical depth of the atmosphere due to atmospheric molecules, u_ω^{cl} is the optical depth of the atmosphere due to water microdroplets of clouds. One can simplify this expression [23] taking in accordance with Fig. 6.9 that radiation from the Earth is absorbed by H_2O and CO_2 molecules at frequencies $\omega < 800\,cm^{-1}$. At larger frequencies absorption of atmospheric molecules gives a small contribution to absorption of the Earth's radiation, i.e. it is determined by water microdroplets of clouds. Assuming in this frequency range u_ω^{cl} is independent of ω, i.e. absorption takes place mostly at a plateau of Fig. 6.20, and taking the distribution function over average optical thicknesses [23] $f(u_\omega^{cl}) \sim \exp(-u_\omega^{cl}/u_o)$, where u_o is an average optical thickness of clouds. Under these assumptions we have [23]

$$u_o = 3.2 \tag{6.91}$$

If we take the average radius of atmospheric water microdroplets with accounting for (3.1) $r = (8 - 12) \, \mu m$, one can find the specific mass of atmospheric condensed water [29] $M = (5 \pm 1) \, mg/cm^2$.

Thus, optical parameters of the atmosphere in the infrared spectrum range allow one to determine typical parameters of atmospheric condensed water for the clear-sky atmosphere. This confirms that the atmosphere consists of regions with various transparency. Cumulus clouds which are responsible for atmospheric electricity are opaque for infrared radiation and are characterized by the optical thickness of the order of $u_o \sim 40$. These clouds include the main part of condensed atmospheric water whose mass is several percent of that of atmospheric water molecules. Radiation from the Earth's surface which passes through the atmosphere takes place for atmospheric regions with rare clouds.

References

1. R.M. Goody, *Atmospheric Radiation: Theoretical Basis* (Oxford University Press, London, 1964)
2. R.M. Goody, Y.L. Yung, *Principles of Atmospheric Physics and Chemistry* (Oxford University Press, New York, 1995)
3. B.M. Smirnov, *Microphysics of Atmospheric Phenomena* (Springer Atmospheric Series, Switzerland, 2017)
4. M. Wendisch, P. Yang, *Theory of Atmospheric Radiative Transfer* (Wiley, Singapore, 2012)
5. M.F. Modest, *Radiative Heat Transfer* (Elsevier, Amsterdam, 2013)
6. K.Y. Kondratyev, *Radiation in the Atmosphere* (Academic Press, New York, 1969)
7. E.J. McCartney, *Absorption and Emission by Atmospheric Gases* (Wiley, New York, 1983)
8. K.N. Liou, *An Introduction to Atmospheric Radiation* (Academic Press, Amsterdam, 2002)
9. G.W. Petry, *A First Course in Atmospheric Radiation* (Sunsign Public School, Madison, 2006)
10. W. Zdunkowski, T. Trautmann, A. Bott, *Radiation in the Atmosphere* (Cambridge University Press, Cambridge, 2007)
11. K.Y. Kondratyev, V.F. Krapivin, C.A. Varotsos, *Global Carbon Cycle and Climate Change* (Springer Praxis Publishing, Chichester, 2003)
12. K.Y. Kondratyev, L.S. Ivlev, V.F. Krapivin, C.A. Varotsos, *Atmospheric Aerosol Properties. Formation, Processes and Impacts* (Springer Praxis Publishing, Chichester, 2006)
13. T.L. Hill, *An Introduction to Statistical Thermodynamics* (Addison Wesley, Reading, MA, 1960)
14. G.N. Lewis, M. Randall, K.S. Pitzer, L. Brewer, *Thermodynamics* (McGraw Hill, New York, 1961)
15. F. Reif, *Statistical and Thermal Physics* (McGrow Hill, Boston, 1965)
16. YaB Zel'dovich, YuP Raizer, *Physics of Shock Waves and High-temperature Hydrodynamic Phenomena* (Academic Press, New York, 1966)
17. G. Kirchhoff, R. Bunsen, Annalen der Physik und Chemie **109**, 275 (1860)
18. A. Beer, Annalen der Physik und Chemie **86**, 78 (1852)
19. J.H. Lambert, *Photometry, or, on the Measure and Gradations of Light, Colors, and Shade* (Eberhardt Klett, Augsburg, 1760)

20. I.I. Sobelman, *Atomic Spectra and Radiative Transitions* (Springer, Berlin, 1979). **Charging of particles**
21. L.D. Landau, E.M. Lifshitz, *Statistical Physics*, vol. 1 (Pergamon Press, Oxford, 1980)
22. V.P. Krainov, H.R. Reiss, B.M. Smirnov, *Radiative Processes in Atomic Physics* (Wiley, New York, 1997)
23. V.P. Krainov, B.M. Smirnov, *Atomic and Molecular Radiative Processes* (Springer Nature, Switzerland, 2019)
24. W. Wien, Wied. Ann. Phys. Chem. **58**, 662 (1896). **Emission of gases**
25. B.M. Smirnov, *Physics of Weakly Ionized Gas* (Nauka, Moscow, 1972; in Russian)
26. B.M. Smirnov, *Physics of Ionized Gases* (Wiley, New York, 2001)
27. B.M. Smirnov, *Physics of Weakly Ionized Gases* (Mir, Moscow, 1980)
28. B.M. Smirnov, JETP **126**, 446 (2018)
29. B.M. Smirnov, *Transport of Infrared Atmospheric Radiation* (Fundamental of the Greenhouse Phenomenon.) (de Gruyter, Berlin, 2020)
30. B.M. Smirnov, EPL **114**, 24005 (2016)
31. https://www.cfa.harvard.edu/
32. http://www.hitran.iao.ru/home
33. E.U. Condon, G.H. Shortley, *The Theory of Atomic Spectra* (Cambridge University Press, Cambridge, 1970)
34. M. Simeckova, D. Jacquemart, L.S. Rothman et al., *JQSRT* 98, 130 (2006)
35. http://www.hitran.org/links/docs/definitions-and-units/
36. L.S. Rothman, I.E. Gordon, Y. Babikov et al., *JQSRT* 130, 4 (2013)
37. I.E. Gordon, L.S. Rothman, C. Hill et al., *JQSRT* 203, 3 (2017)
38. G. Herzberg, *Molecular Spectra and Molecular Structure* (Van Nostrand Reinhold, Princeton, 1945)
39. G.M. Barrow, *Introduction to Molecular Spectroscopy* (McGraw-Hill, New York, 1962)
40. H.C. Allen, P.C. Cross, *Molecular Vibrators: The Theory and Interpretation of High Resolution Infrared Spectra* (Wiley, New York, 1963)
41. M.A. El'yashevich, *Molecular Spectroscopy* (Fizmatgiz, Moscow, 1963; in Russian)
42. J.I. Steinfeld, *Molecules and Radiation* (Dover, New York, 1985)
43. S. Svanberg, *Atomic and Molecular Spectroscopy* (Springer, Berlin, 1991)
44. C. Banwell, E. McCash, *Fundamentals for Molecular Spectroscopy* (McGrow Hill, London, 1994)
45. P.S. Sindhu, *Fundamentals of Molecular Spectroscopy* (New Age International, Dehli, 2006)
46. S. Chandra, *Molecular Spectroscopy* (Alpha Science International, Dehli, 2009)
47. J.L. McHale, *Molecular Spectroscopy* (CRC Press, Boca Raton, 2017)
48. V.P. Krainov, B.M. Smirnov, JETP **129**, 9 (2019)
49. http://www1.lsbu.ac.uk/water/water-vibrational-spectrum
50. A.A. Radzig, B.M. Smirnov, *Reference Data on Atoms, Molecules, and Ions* (Springer, Berlin, 1985)
51. S.V. Khristenko, A.I. Maslov, V.P. Shevelko, *Molecules and Their Spectroscopic Properties* (Springer, Berlin, 1998)
52. G. Mie, *Annalen der Physik* **330**, 377 (1908). §2
53. J.A. Stratton, *Electromagnetic Theory* (McGraw-Hill, New York, 1941)
54. H.C. van de Hulst, *Light Scattering by Small Particles* (Wiley, New York, 1957)
55. C.F. Bohren, D.R. Huffmann, *Absorption and Scattering of Light by Small Particles* (Wiley, New York, 2010)
56. https://en.wikipedia.org/wiki/Properties-of-water
57. T.S. Light et al., Electrochem. Solid State Lett. **8**, E16 (2005)
58. L.D. Landau, E.M. Lifshitz, *Electrodynamics of Continuous Media* (Pergamon Press, Oxford, 1984)
59. B.M. Smirnov, *Clusters and Small Particles in Gases and Plasmas* (Springer NY, New York, 1999)
60. https://en.wikipedia.org/wiki/Refractive-index

61. H.D. Downing, D.W. Williams, J. Geoph. Res. **80**, 1656 (1975)
62. D.A. Draegert, N.W.B. Stone, B. Curnutte, D. Williams, J. Opt. Soc. Am. **56**, 64 (1966)
63. W.M. Irvine, J.B. Pollack, Icarus **8**, 324 (1968)
64. M.R. Querry, B. Curnutte, D. Williams, J. Opt. Soc. Am. **59**, 1299 (1969)
65. V.M. Zolatarev, B.A. Mikhailov, L.I. Aperovich, S.I. Popov, Opt. Spectrosc. **27**, 430 (1969)
66. C.W. Robertson, D. Williams, J. Opt. Soc. Am. **61**, 1316 (1971)
67. A.N. Rusk, D. Williams, M.R. Querry, J. Opt. Soc. Am. **61**, 895 (1971)
68. P.S. Ray, Appl. Opt. **11**, 836 (1972)
69. C.W. Robertson, B. Curnutte, D. Williams, Mol. Phys. **26**, 183 (1973)
70. O. Boucher, *Atmospheric Aerosols. Properties and Climate Impacts* (Springer, Dordrecht, 2015)
71. D. Eisenberg, W. Kauzmann, *The Structure and Properties of Water* (Oxford University Press, New York, 1969)
72. J.B. Hasted, *Aqueous Dielectrics* (Chapman and Hall, London, 1973)
73. M.N. Afsar, J.B. Hasted, Infrared Phys. **18**, 835 (1978)
74. https://en.wikipedia.org/wiki/Electromagnetic-absorption-by-water
75. http://www1.lsbu.ac.uk/water/water-vibrational-spectrum.html
76. C.M.R. Platt, Quart. J. R. Meteorol. Soc. **102**, 553 (1976)
77. S. Twomey, Geofis. Pure Appl. **43**, 227 (1959)
78. S. Twomey, J. Atmos. Sci. **34**, 1149 (1977)
79. A.V. Gurevich, E.E. Tsedilina, *Long Distance Propagation of HF Radio Waves* (Springer, Berlin, 1985)
80. J.S. Seybold, *Introduction to RF Propagation* (Wiley, Hoboken, New Jersey, 2005)
81. https://en.wikipedia.org/wiki/Radio-propagation
82. H.J. Liebe, G.A. Hufford, T. Manabe, Int. J. Infrared Millim. Waves. **12**, 659 (1991)
83. M.L. Salby, *Physics of the Atmosphere and Climate* (Cambridge University Press, Cambridge, 2012)
84. D.L. Hartmann, *Global Physical Climatology* (Elesevier, Amsterdam, 2016)
85. B.M. Smirnov, J. Phys. D Appl. Phys. **51**, 214004 (2018)
86. Palaeosens Project Members, Nature **491**, 683 (2012)
87. L.B. Stap, P. Köhler, G. Lohmann, Earth Syst. Dynam. **10**, 333–345 (2019)
88. J. Feichter, E. Roeckner, U. Lohmann, B. Liepert, J. Clim. **17**, 2384 (2004)
89. J. Hansen et al., *J. Geophys. Res.* **110**, D18104 (2005)
90. https://en.wikipedia.org/wiki/Climate-sensitivity
91. Intergovernmental Panel on Climate Change. *Nature* **501**, 297–298 (2013). http://www.ipcc.ch/pdf/assessment?report/ar5/wg1/WGIAR5-SPM-brochure-en.pdf
92. J.T. Fasullo, K.E. Trenberth, Science **338**, 792 (2012)
93. G.N. Plass, Tellus VIII **141** (1956)
94. N. Andronova, M.E. Schlesinger, J. Geophys. Res. **106**, D22605 (2001)
95. M.A. Snyder, J.L. Bell, L.C. Sloan, Geophys. Res. Lett. **29**, 014431 (2002)
96. J.D. Annan, J.C. Hargreaves, Geophys. Res. Lett. **33**, L06704 (2006)
97. A. Ganopolski, T. Schneider von Deimling, Geophys. Res. Lett. **35**, L23703 (2008)
98. M.E. Walter, Not. Am. Mat. Soc. **57**, 1278 (2010)
99. A. Schmittner, N.M. Urban, J.D. Shakun, *Science* **334**, 1385 (2011)
100. *The Warming Papers*, ed. by D. Archer, R. Pierrehumbert (Wiley-Blackwill, Oxford, 2011)
101. S. Arrhenius, Phil. Mag. **41**, 237 (1896)
102. G.S. Calendar, Weather **4**, 310 (1949)
103. G.N. Plass, D.I. Fivel, Quant. J. Roy. Met. Soc. **81**, 48 (1956)
104. W.M. Elsasser, Phys. Rev. **54**, 126 (1938)
105. B.M. Smirnov, Int. Rev. At. Mol. Phys. **10**, 39 (2019)
106. B.M. Smirnov, J. Atmos. Sci. Res. **2**, N4, 21 (2019)
107. M. Milankovich, *Theorie Mathematique des Phenomenes Thermiques produits par la Radiation Solaire* (Gauthier-Villars, Paris, 1920)
108. M. Milankovich, *Canon of Insolation and the Ice Age Problem* (Belgrade, 1941)

109. *U.S. Standard Atmosphere* (Washington, U.S. Government Printing Office, 1976)
110. https://en.wikipedia.org/wiki/Outgoing-longwave-radiation
111. R.T. Pierrehumbert, *Principles of Planetary Climate* (Cambridge University Press, New York, 2010)
112. R.T. Pierrehumbert, Phys. Today **64**, 33 (2011)
113. W. Zhong, J.D. Haigh, Weather **68**, 100 (2013)

Chapter 7
Conclusion

The goal of the above analysis was to combine the contemporary physical picture of global atmospheric phenomena involving water with information about elementary processes in these phenomena. As a result, one can understand deeper some aspects of these phenomena. In this consideration, we use models which contain the basic features of processes under consideration, including the global character of the phenomena, and hence, this physical picture remains qualitative. Below we formulate the understanding of the global atmospheric phenomena involving water processes with accounting for the above analysis.

The phenomenon of water circulation through the atmosphere remains the same as it was formulated decades ago. Namely, water evaporates from the Earth's surface in the form of water molecules. An ascending flux of water molecules rises up to cold layers of the atmosphere where condensation of this vapor proceeds. Then condensed water in the form of rain drops or snow falls down toward the Earth's surface, and in this manner the balance is realized for evaporated water molecules and that attained the Earth's surface in the form of raindrops.

Accepting this scheme in principle, one can indicate some details which influence our understanding of this problem. Indeed, air transport in the atmosphere results in the form of convection, and typical sizes of vortices in this process are found roughly in the range (30–100) cm. Hence, transport of air for large distance has the diffusion character, and returning to the Earth proceeds through the drift motion. If water evaporates from the Earth's surface, their molecules are captured by air vortices and they together partake in the diffusion and drift motion. Correspondingly, the drift channel of water returning to the Earth is possible.

But along with this character of motion of water molecules, they are lost at some altitudes as a result of condensation. This process influences on the distribution function of water molecules over altitudes which differs from that of air molecules

© The Editor(s) (if applicable) and The Author(s), under exclusive license
to Springer Nature Switzerland AG 2020
B. M. Smirnov, *Global Atmospheric Phenomena Involving Water*,
Springer Atmospheric Sciences, https://doi.org/10.1007/978-3-030-58039-1_7

because of condensation. Hence, the altitude profile of the number density of water molecules which is determined reliably in balloon measurements contains information about the condensation process. In particular, the altitude profile of molecular water at a given locality allows one to determine which part of the water flux is connected with condensed water. Rough approximations in the above analysis show that the condensed part of the water flux to the Earth exceeds that due to water molecules by several times. This confirms that the classical scheme of water circulation in the atmosphere holds true in principle.

One can expect the validity of the classical explanation of the character of water condensation in the atmosphere. According to this explanation, the condensation process proceeds at some higher altitudes because the temperature is low here. But according to the model of standard atmosphere which is based on the convection character of air motion, the average partial pressure of a water vapor is below the saturated pressure of a water vapor at which the condensation process starts. Moreover, the average air moisture, i.e. the ratio of these pressures, decreases with an increasing altitude. This means that water condensation is impossible in the atmosphere with average parameters, i.e. condensation results from atmospheric fluctuations. In other words, formation of the condensed phase of water in the atmosphere proceeds under nonequilibrium conditions in the atmosphere.

According to the above analysis, water condensation in the atmosphere results under the action of vertical winds which are absent in the model of standard atmosphere, i.e. in the atmosphere with average parameters. Because it is determined by a divergence of atmospheric parameters from average ones, the amount of condensed water in the atmosphere is small compared to that of water molecules. The condensation process in the atmosphere results from mixing of wet warm air from lower layers with cold ones located at larger altitudes, so that supersaturated water is formed in restricted air regions. As a result, condensed water exists only in a small part of the atmosphere. Note that the total density of water in these regions exceeds that corresponded to the saturated vapor pressure because the equilibrium between free and bonded water molecules establishes fast compared with a time of water evolution in the atmosphere. As a result of this equilibrium, the number density of free water molecules is equal to that at the saturated vapor pressure for the temperature of a given region.

The analysis of nucleation processes, i.e. the growth process of the condensed phase of water in the atmosphere, allows one to understand, what is the condensed phase. At the first stage of condensation after mixing of air layers (parcels), a water vapor becomes supersaturated. Then nanoclusters of water are formed as a result of attachment of water molecules to nuclei of condensation which are atmospheric ions and radicals. This stage proceeds relatively fast and its time depends on the density of condensation nuclei. As a result of the growth stage, the partial pressure of a water vapor in the form of free water molecules establishes to be equal to that at the saturated vapor pressure for this air temperature.

The next stage of growth of water nanoclusters or droplets, i.e. the water condensed phase, proceeds due to interaction between a water vapor and condensed phase through processes of evaporation of free molecules from the droplet surface and attachment of molecules to their surface. For simplicity, we restrict by the liquid state of the water condensed phase, and this is in the form of small droplets located in atmospheric air. Because of the equilibrium, the average number of acts per unit time for molecule attachment to droplet surfaces is equal to that of molecule evaporation. But for an individual droplet, this equilibrium may be violated because of a different dependence of evaporation and attachment rates on the droplet size. As a result, small droplets evaporate, whereas large droplets grow, and the average droplet size increases in time. This corresponds to the coalescence mechanism of droplet growth or the mechanism of Ostwald ripening. This growth mechanism is realized at the first growth stage of droplet growth in supersaturated air.

The rate of coalescence decreases with an increasing droplet size, and starting from certain sizes, the gravitation mechanism of droplet growth dominates. In this mechanism, large droplets join with small ones in the course of their falling down under the action of their weight, since the falling velocity for large droplets is higher than that for small ones. The growth rate for this mechanism increases strongly with an increasing droplet size. For a typical water density in cumulus clouds which contain the most part of condensed atmospheric water, this process of transformation of small droplets into raindrops lasts of the order of ten minutes if droplets are neutral. This time is small compared with a typical lifetime of cumulus clouds. From this contradiction follows that joining water droplets are charged, and their charge has the same sign. This means that growth processes in clouds proceed simultaneously with the processes of droplet charging, i.e. processes of atmospheric electricity accompany processes of droplet growth. These processes of growth and charging of water droplets proceed in cumulus clouds and become slower with an increasing droplet size. From this, it follows that cumulus clouds contain charged droplets of a micrometer size.

There are two mechanisms of charging of water droplets in the atmosphere. In the first case, collision of two neutral droplets which are found in different aggregate states may lead to transfer of a charge from one particle to another one. Indeed, water droplets may be found in different aggregate states, i.e. they can be liquid, solid (ice particles), they can have an amorphous solid structure (snow particle or hailstone) or be a mixture of different aggregate states, as graupel particles consisting of the mixture of liquid water and snow. Because water is a weak electrolyte that contained ions H_3O^+ and OH^- in water, electrolyte properties may depend on the aggregate state. Therefore, a contact of two water particles in different aggregate states leads to transition of some ions from one particle to another one as a result of different contact potentials for these particles. Hence, collision of two water neutral particles in different aggregate state leads to charging of colliding particles.

Another mechanism of charging of water droplets in the atmosphere results from attachment of atmospheric ions to liquid water droplets. Indeed, atmospheric ions are formed in the atmosphere under the action of cosmic rays and decay as a result of recombination of positive and negative ions. If liquid droplets are located in the atmosphere, an additional channel of ion loss occurs as a result of attachment of negative and positive atmospheric ions to water droplets. The drift velocity of ions in atmospheric air is determined by ion mobilities, and if they are different for negative and positive atmospheric ions, the droplet acquires a certain charge which equalizes currents of negative and positive ions to its surface. As a result, liquid droplets obtain charges which are determined by the mobility difference for negative and positive ions.

We have that water microdroplets in a cumulus cloud may be charged negatively or positively depending on types of ions. In turn, sorts of ions in air are determined by admixtures. But though microdroplets may have any sign, the Earth is charged negatively, and hence the negative charge of microdroplets is realized more often. This is of importance for atmospheric electricity. According to the model of global electric circuit which exists a century, the Earth is a spherical capacitor, and its negative electrode is the Earth's surface, while the ionosphere as a region of high conductivity is used often as the second electrode. In reality, the main contribution to the resistance between electrodes of this capacitor follows from the troposphere, and the tropopause may be used as the second electrode of global electric circuit. Then the electric current of this circuit results from motion of tropospheric ions under the action of the Earth's electric field that leads to the Earth's discharging.

In this consideration, we divide the atmosphere over the Earth's surface into two parts. The first one corresponds to a clear sky where charged microdroplets or aerosols are absent practically. This atmospheric part is just included in the global electric circuit, and the conductivity of this atmosphere is determined by ions formed under the action of cosmic rays. Another atmospheric part which is responsible for the Earth's charging contains cumulus clouds. This atmosphere part includes the most part of atmospheric condensed water mostly in the form of water microdroplets which move to the Earth under the action of gravitation forces. The negative electric current resulted from falling down these charged particles charges the Earth negatively.

The charging process may be realized in two ways. In the first case, charged water microdroplets attain the Earth directly and can cause corona discharge near the Earth's surface in the vicinity of some objects on the Earth surface. In the second case which is realized more seldom, water microdroplets lost the charge at some altitudes, and then the subsequent partial discharging of clouds may be realized through lightning. Because the charge loss for water microdroplets depends on the air temperature, lightning may be realized in a warm season. In addition, a release of the charge from water microdroplets causes their fast joining which was restricted early by repulsion of charged microdroplets. Therefore, lightnings are accompanied usually with rain.

In reality, the first charging mechanism as a result of collisions is responsible for charging of flights or cosmic bodies as a result of collision with water microparticles. The second charging mechanism due to falling down of charged water microdroplets is responsible for the Earth's charging. This charging mechanism requires more soft conditions and may be realized in a wide range of atmospheric altitudes. In addition, the second charging mechanism through charged water microparticles leads to separation of atmospheric charges because falling down of charged microparticles creates an electric current. We give in Fig. 7.1 the basic processes of atmospheric electricity as the secondary phenomenon of water circulation through the atmosphere. The main contribution to the Earth's charging follows from regions of the Earth's surface covered by cumulus clouds. Discharging of the Earth proceeds in atmospheric

Fig. 7.1 Basic processes of water circulation (wide arrows) and charge transport (fine arrows with directions of negative charge transfer) in the atmosphere

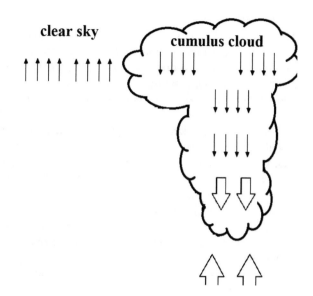

regions of clear sky as a result of drift of molecular ions formed under the action of cosmic rays.

It should be noted that dividing the atmosphere into two parts, with and without water microdroplets, testifies also about two regimes of equilibrium of atmospheric water and that located at the Earth's surface. In the first regime of water residence in the atmosphere, water molecules return to the Earth's surface as a result of drift together with air molecules. In the second regime, they attach to water droplets, and water returns to the Earth as a result of falling down of water microdroplets or rain, if microdroplets lose the charge and join in raindrops. As is seen, this process depends on the charge of water droplets. Hence, processes of atmospheric electricity which are secondary ones with respect to circulation atmospheric water, also influence on the rate of water processes. The Earth's water is a reservoir for atmospheric water, and the equilibrium between them is governed by processes of evaporation of water molecules from the Earth's surface and their attachment to the surface.

One can see that atmospheric electricity is determined by atmospheric condensed water. In turn, the amount of condensed water in the atmosphere is small compared with that due to water molecules because formation of the water condensed phases results from nonequilibrium processes. From the analysis of atmospheric processes, one can estimate the part of condensed water in the atmosphere that is $\sim 10\%$, and practically all condensed water is located in cumulus clouds. But condensed water is of importance also for thermal emission of the atmosphere. Indeed, approximately 30% of the radiative flux from the atmosphere to the Earth is created by water microdroplets, and the contribution of atmospheric condensed water to outgoing thermal emission of the atmosphere is estimated as $\sim 50\%$. Therefore, one can expect that the analysis of thermal emission of the atmosphere allows one to determine more precisely the amount of condensed atmospheric water.

One can expect that combining electric and radiative atmospheric processes, one can determine more accurate the condensed part of atmospheric water. But in reality this is not so, because atmospheric emission is determined by the globe regions with a low optical thickness, whereas the Earth's charging is connected with cumulus clouds which cover a small part of the Earth's surface ($\sim$10%). Therefore, the main contribution to emission of atmospheric condensed water is from rare clouds. The amount of this condensed water in cumulus clouds is estimated as $\sim$(1–10)% of the total water amount in the atmosphere. In regions of residence of cumulus clouds, the atmospheric optical thickness with respect to condensed water is 10–100 times higher than that in atmospheric regions with rare clouds which give the main part of infrared emission of the atmosphere due to water condensed phase.

One more peculiarity of this analysis is connected with simple models and numerical data which are used in the analysis of atmospheric phenomena involving water. In this analysis, for simplicity, we deal with the average values of parameters instead of the distribution function over them. For example, we use a certain size and charge of water microdroplets in cumulus clouds which are identical for all the droplets, whereas it is more correct to be based on the size and charge distribution functions for these quantities. This leads to the obviousness of the analysis, but the results are estimations. This approach allows us to choose appropriate models for objects and processes under consideration, but the yield physical picture has the qualitative character. The basic advantage of this approach is that on the basis of numerical evaluations one can refuse from concepts which results differ in orders of magnitude from observed or measured ones.

One can expect that physics of atmospheric electricity is universal for all planets, but our experience shows another situation. Moreover, as it follows from this analysis, in the Earth's atmosphere this phenomenon results from processes involving free and bonded water molecules, whereas in atmospheres of other planets water is absent in appropriate amount. In this consideration, transport of large air masses leaves aside, but just these processes create atmospheric nonuniformities which are the basis of condensation processes.

Appendix
Appendices

A.1 Fundamental Units

Table A.1 Basic fundamental physical constants

Electron mass	$m_e = 9.10938 \times 10^{-28}$ g
Proton mass	$m_p = 1.67262 \times 10^{-24}$ g
Atomic unit of mass	$m_a = \frac{1}{12} m(^{12}C) = 1.660539 \times 10^{-24}$ g
Ratio of proton and electron masses	$m_p/m_e = 1836.15$
Ratio of atomic and electron masses	$m_a/m_e = 1822.89$
Electron charge	$e = 1.602177 \times 10^{-19} C = 4.803204 \times 10^{-10} e.s.u.$
	$e^2 = 2.3071 \times 10^{-19}$ erg $\cdot$ cm $= 14.3996$ eV$\cdot$Å
Planck constant	$h = 6.62619 \times 10^{-27}$ erg $\cdot$ s
	$\hbar = 1.054572 \times 10^{-27}$ erg $\cdot$ s
Light velocity	$c = 2.99792458 \times 10^{10}$ cm/s, $m_e c^2 = 510.98$ keV
Fine-structure constant	$\alpha = e^2/(\hbar \cdot c) = 0.07295$
Inverse fine-structure constant	$1/\alpha = \hbar \cdot c/e^2 = 137.03599$
Bohr radius	$a_o = \hbar^2/(m_e \cdot e^2) = 0.5291772$Å
Rydberg constant	$R = m_e e^4/(2\hbar^2) = 13.60569$ eV $= 2.17987 \times 10^{-18}$ J
Bohr magnetron	$\mu_B = e\hbar/(2m_e c) = 9.27401 \times 10^{-24} J/T = 9.27401 \times 10^{-21}$ erg/Gs
Avogadro number	$N_A = 6.02214 \times 10^{23}$ mol^{-1}
Stefan–Boltzmann constant	$\sigma = \pi^2/(60\hbar^3 c^2) = 5.670374 \times 10^{-12}$ W/(cm$^2 \cdot$ K^4)
Molar volume	$R = 22.414$ l/mol
Loschmidt number	$L = N_A/R = 2.6867 \times 10^{19}$ cm^{-3}
Faraday constant	$F = N_A e = 96485.3$ C/mol

© The Editor(s) (if applicable) and The Author(s), under exclusive license
to Springer Nature Switzerland AG 2020
B. M. Smirnov, *Global Atmospheric Phenomena Involving Water*,
Springer Atmospheric Sciences, https://doi.org/10.1007/978-3-030-58039-1

A.2 Conversional Factors for Units

A.2.1 Conversional Factors for Energy Units

Table A.2 Conversional factors for units of energy

	1 J	1 erg	1 eV	1 K	1 cm^{-1}	1 MHz	1 kcal/mol	1 kJ/mol
1 J	1	10^7	6.242×10^{18}	7.243×10^{22}	5.034×10^{22}	1.509×10^{27}	1.439×10^{20}	6.022×10^{20}
1 erg	10^{-7}	1	6.242×10^{11}	7.243×10^{15}	5.034×10^{15}	1.509×10^{20}	1.439×10^{13}	6.022×10^{13}
1 eV	1.602×10^{-19}	1.602×10^{-12}	1	11604	8065.5	2.418×10^8	23.045	96.485
1 K	1.381×10^{-23}	1.381×10^{-16}	8.617×10^{-5}	1	0.69504	2.084×10^4	1.987×10^{-3}	8.314×10^{-3}
1 cm^{-1}	1.986×10^{-23}	1.986×10^{-16}	$1.2398 \cdot 10^{-4}$	1.4388	1	2.998×10^4	2.859×10^{-3}	1.196×10^{-2}
1 MHz	6.626×10^{-28}	6.626×10^{-21}	4.136×10^{-9}	4.799×10^{-5}	3.336×10^{-5}	1	9.537×10^{-9}	3.9903×10^{-7}
1 kcal/mol	6.948×10^{-21}	6.948×10^{-28}	$4.336 \cdot 10^{-2}$	503.22	349.76	1.048×10^7	1	4.184
1kJ/mol	1.660×10^{-21}	1.660×10^{-28}	1.036×10^{-2}	120.27	83.594	$2.506 \cdot 10^6$	0.2390	1

A.2.2 Conversional Factors for Charge Units

Table A.3 Units of electric charge

	1 e	1 $CGSE$	1 C
1 e	1	4.8032×10^{-10}	1.60218×10^{-19}
1 $CGSE$	2.0819×10^9	1	3.33564×10^{-10}
1 C	6.2415×10^{18}	2.99792×10^9	1

A.2.3 Conversional Factors for Conductivity Units

Table A.4 Units of conductivity

	S/m	$1/(\Omega \cdot cm)$	$1/s$
S/m	1	0.01	8.98755×10^9
$1/(\Omega \cdot cm)$	100	1	8.98755×10^{11}
$1/s$	1.11265×10^{-10}	1.11265×10^{-12}	1

A.2.4 Conversional Factors for Current Density Units

Table A.5 Units of current density

	1 e/(cm$^2 \cdot$ s)	1 $CGSE$	1 A/m^2
1 e/(cm$^2 \cdot$ s)	1	2.99792 4.8032×10^{-10}	1.60218×10^{-15}
1 $CGSE$	2.0819×10^9	1	3.3356×10^{-6}
1 A/m^2	6.2415×10^{14}	2.9979×10^5	1

A.2.5 Conversional Factors for Viscosity Units

Table A.6 Units of viscosity

	1 $CGSE = $ g/(cm $\cdot$ s)	1 P (poise)	1 $Pa \cdot s$
1 $CGSE = $ g/(cm $\cdot$ s)	1	1	0.1
1 P (poise)	1	1	0.1
1 $Pa \cdot s$	10	10	1

A.3   Conversional Factors for Physical Formulas

Table A.7 Physical formulas in different units

Number	Formula [a]	Factor C	Units used
1.	$v = C\sqrt{\varepsilon/m}$	5.931×10^7 cm/s	ε in eV, m in $e.m.u.$ [a]
		1.389×10^6 cm/s	ε in eV, m in $a.m.u.$ [a]
		5.506×10^5 cm/s	ε in K, m in e.m.u.
		1.289×10^4 cm/s	ε in K, m in a.m.u.
2.	$v = C\sqrt{T/m}$	1.567×10^6 cm/s	T in eV, m in $a.m.u.$
		1.455×10^4 cm/s	ε in k, m in $a.m.u.$
3.	$\varepsilon = Cv^2$	$3.299 * 10^{-12}$ K	v in cm/s, m in $e.m.u.$
		6.014×10^{-9} K	v in cm/s, m in $a.m.u.$
		$2.843 * 10^{-16}$ eV	v in cm/s, m in $e.m.u.$
		5.182×10^{-13} eV	v in cm/s, m in $a.m.u.$
4.	$\omega = C\varepsilon$	$1.519 \times 10^{15} s^{-1}$	ε in eV
		$1.309 \times 10^{11} s^{-1}$	ε in k
5.	$\omega = C/\lambda$	1.884×10^{15} s^{-1}	λ in μm
6.	$\varepsilon = C/\lambda$	1.2398 eV	λ in μm
7.	$\omega_H = CH/m$	$1.759 \times 10^7 s^{-1}$	H in Gs, m in $e.m.u.$
		$9655 s^{-1}$	H in Gs, m in $a.m.u.$
8.	$r_H = C\sqrt{\varepsilon m}/H$	3.372 cm	ε in eV, m in $e.m.u.$, H in Gs
		143.9 cm	ε in eV, m in $a.m.u.$, H in Gs
		3.128×10^{-2} cm	ε in K, m in $e.m.u.$, H in Gs
		1.336 cm	ε in K, m in $a.m.u.$, H in Gs
9.	$p = CH^2$	$4.000 \times 10^{-3} Pa =$ 0.04 erg/cm^3	H in Gs

[a] $e.m.u.$ is the electron mass unit ($m_e = 9.108 \times 10^{-28}$ g), $a.m.u.$ is the atomic mass unit ($m_a = 1.6605 \times 10^{-24}$ g)

Explanations to Table A.7.

1. The particle velocity is $v = \sqrt{2\varepsilon/m}$, where ε is the energy, m is the particle mass.
2. The average particle velocity is $v = \sqrt{8T/(\pi m)}$ with the Maxwell distribution function of particles on velocities; T is the temperature expressed in energetic units, m is the particle mass.
3. The particle energy is $\varepsilon = mv^2/2$, where m is the particle mass, v is the particle velocity.
4. The photon frequency is $\omega = \varepsilon/\hbar$, where ε is the photon energy.
5. The photon frequency is $\omega = 2\pi c/\lambda$, where λ is the wavelength.
6. The photon energy is $\varepsilon = 2\pi\hbar c/\lambda$.
7. The Larmor frequency is $\omega_H = eH/(mc)$ for a charged particle of a mass m in a magnetic field of a strength H.

8. The Larmor radius of a charged particle is $r_H = \sqrt{2\varepsilon/m}/\omega_H$, where ε is the energy of a charged particle, m is its mass, ω_H is the Larmor frequency.

9. The magnetic pressure $p_m = H^2/(8\pi)$.

Index

© The Editor(s) (if applicable) and The Author(s), under exclusive license
to Springer Nature Switzerland AG 2020
B. M. Smirnov, *Global Atmospheric Phenomena Involving Water*,
Springer Atmospheric Sciences, https://doi.org/10.1007/978-3-030-58039-1

Printed in the United States
by Baker & Taylor Publisher Services